BASIC CONCEPT

OF

STEREOCHEMISTRY

By:

Dr. HARDIK. B. BHATT

I/C PRINCIPAL,

DEPARTMENT OF CHEMISTRY,

GYANMAJARI INSTITUTE,

BHAVNAGAR- 360020 (INDIA)

Dr. BHAVIN B DHADUK

ASSISTANT PROFESSOR,

DEPARTMENT OF CHEMISTRY,

SCHOOL OF SCIENCE,

RK UNIVERSITY,

RAJKOT- 360020 (INDIA)

ABOUT AUTHOR

Dr. Hardik B Bhatt

Dr. Hardik B Bhatt is I/C Principal in Gyanmanjari Science College. He has got 12 years of teaching experience including teaching at Shree N M G P Institute, Ranpur and R. K. University, Rajkot. He was the organizing committee member in a National conference NCRIS-2018 held at R. K. University. Dr. Bhatt has published & presented 10 research papers in journal of national and international repute. He also wrote a book Engineering Chemistry-I. He has done research in the field of newer photochemical estimation method and was awarded the degree of Ph.D. in year 2014. He also guided 6 student of M.Sc. organic chemistry for their dissertation work and 1 research scholar.

Dr. Bhavin Dhaduk

Dr. Bhavin Dhaduk received his Master Degree (2011) and Ph.D. Degree (2016) in Analytical Pharmaceutical Chemistry from Saurashtra University, Rajkot (Gujarat- India). After masters he worked in INTAS Pharmaceutical LTD., Ahmedabad as a Research Associate and after that he joined his Post-Doctoral study. His research work mainly focused on synthesis, method development and validation, ultrasonic, thermal and XRD studies of derivatives of bisphenols compounds. He is the author of more than 12 national and international journal articles. He worked as Assistant Professor in Shree H. N. Shukla Science College, Rajkot in 2015. Now he is working as an Assistant Professor in School of Science, RK University in Rajkot. He also guided 20 student of M.Sc. Analytical chemistry for their dissertation work and 1 research scholar.

ABOUT THIS BOOK

Molecular shape, form and symmetry play a central role in organic chemistry. The aim of this book is to offer a decent understanding of conceptual basis of stereochemistry. Mainly focus lies in the fundamentals of structural stereochemistry rather than the dynamic aspects that are more relevant to reaction mechanisms. In this book, we discuss the basic principles, conformations and configurations, the methods for writing structures in two dimensional and three dimensional projections and their interconversions and chirality. It also discusses the dependence of optical activity on structure and concludes with an examination of topological isomerism. This book is written especially for the students at undergraduate and postgraduate level.

KEY FEATURES

The book includes problems and solutions at the end.

TABLE OF CONTENT

Basic terms/ Stereoisomerism/ Conformation isomerism/ Conformations of ethane/ Torsional energy/ Conformations of butane/ Conformations of cyclohexane/ Conformations of monosubsituted cyclohexane derivatives/ Magnitude of 1,3-diaxial interaction/ Disubsituted cyclohexane/ Conformational effects on stability/ Geometrical isomerism/ E-Z systems of nomenclature/ Optical activity/ Optical isomerism/ Racemic mixture/ R-S Nomenclature/ Solved problems/ Question asked in SET examinations with answer.

AUDIENCE

Undergraduate and Post graduate students

BASIC CONCEPT OF STEREOCHEMISTRY

BASIC TERMS

1) **ANTI CONFORMATION** :

   For example, anti-conformation of butane, two methyl groups at an angle of 180° to each other.

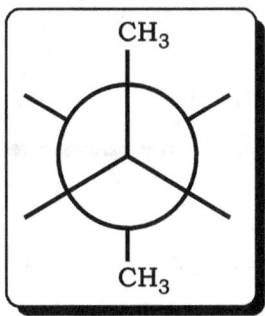

2) **CHAIR CONFORMATION** :

   Staggered conformation of cyclohexane that has no angle strain or torsional strain and is therefore lowest energy conformation.

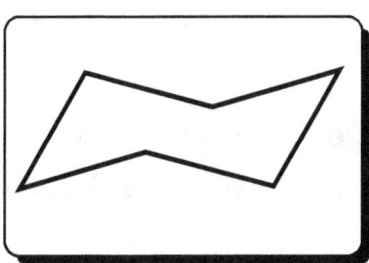 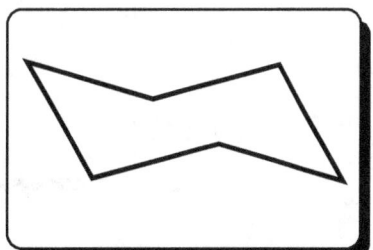

3) **CHIRALITY** :

   The property of having handedness.

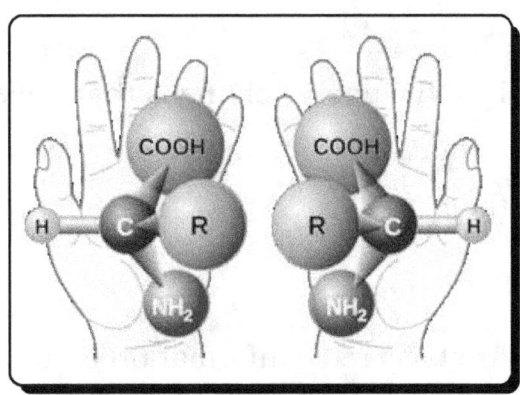

4) CHIRAL MOLECULE :

A molecule that is not superimposable on its mirror image. Chiral molecules have handedness and are capable of existing as a pair of enantiomers.

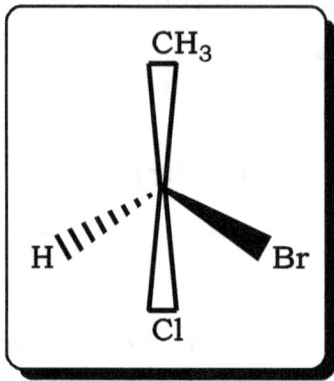

5) CONFIGURATION :

The particular arrangement of atoms of groups in space that is characteristic of a given stereoisomer.

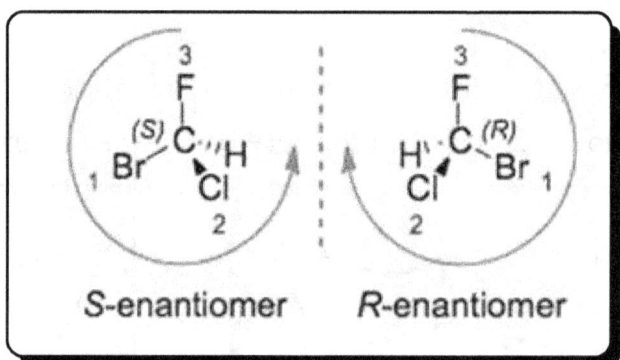

S-enantiomer   R-enantiomer

6) CONFORMATION :

A particular temporary orientation of a molecule that results from rotations about its single bond.

7) CONFORMATIONAL ANALYSIS :

An analysis of the energy changes that molecule undergoes as its groups undergo rotation (sometimes only partial) about the single bonds that join them.

8) CONFORMER :

A particular staggered conformation of a molecule.

9) DEXTROROTATORY :

A compound that rotates plane polarized light clockwise.

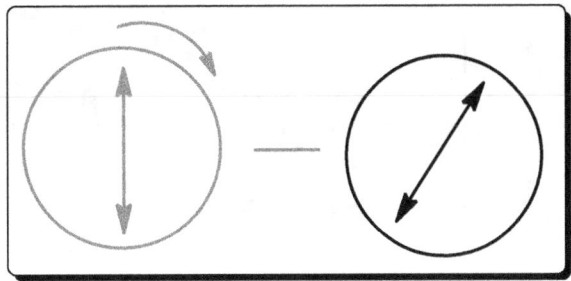

10) DISTEREOMERS :

Stereoisomers that are not mirror image of each other.

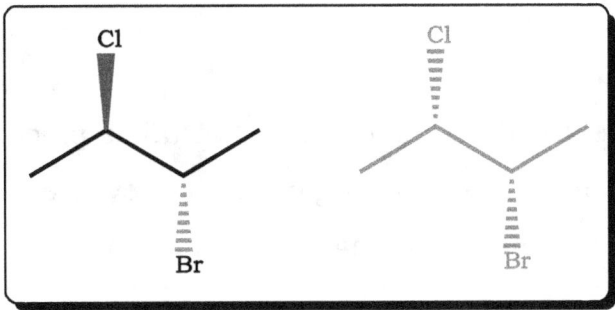

11) ECLIPSED CONFORMATION :

A temporary orientation of groups around two atoms joined by a single bond such that the groups directly oppose each other.

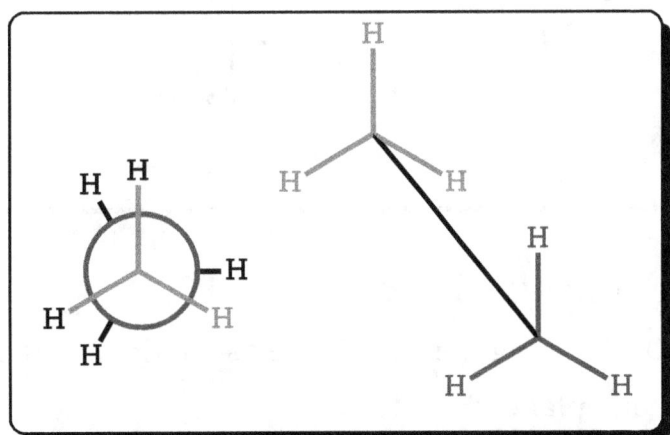

12) GAUCHE CONFORMATION :

A gauche conformation of butane, for example, in which methyl groups at an angle of 60° to each other.

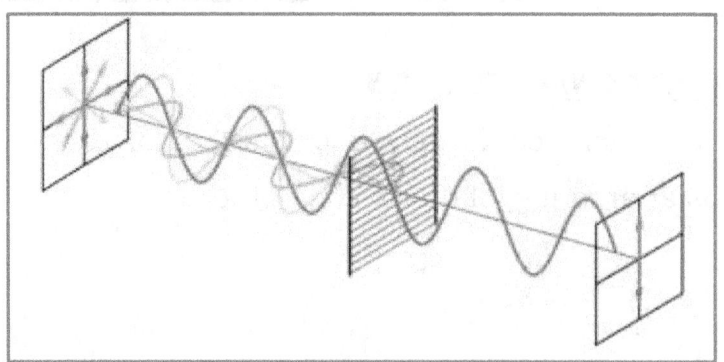

13) ISOMERS :

Different molecules that have the same molecular formula.

14) MUTAROTATION :

The spontaneous change that takes place in the optical rotation of α & β anomers of a sugar when they are dissolved in water. The optical rotations of the sugars change until they reach the same value.

$$[\alpha] = +52.7^0$$

15) PLANE POLARISED LIGHT :

Ordinary light in which the oscillations of the electrical field occur only in one plane.

16) RACEMIC FORM (RECEMIC MIXTURE) :

An equimolar mixture of enantiomers. A racemic form is optically inactive.

17) MESO COMPOUND :

An optically inactive compound having a multiple stereocenters that is superimposable on its mirror image.

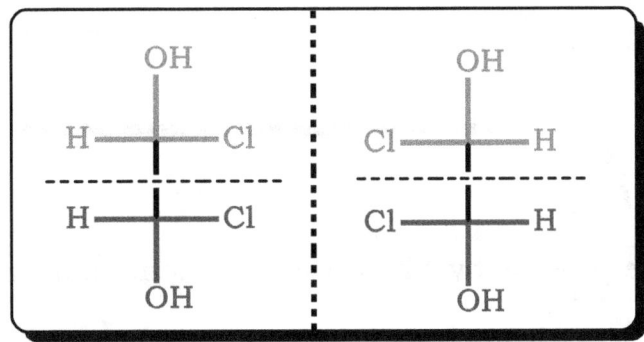

18) OPTICALLY ACTIVE COMPOUNDS :

A compound that rotates the plane of polarization of plane polarized light.

19) PERIPLANER :

A conformation in which vicinal groups lie in the same plane.

20) STAGGERED CONFORMATION :

A temporary orientation of groups around two atoms joined by a single bond such that the bonds of the back atom exactly bisect the angles formed by the bonds of the front atom in Newman projection formula.

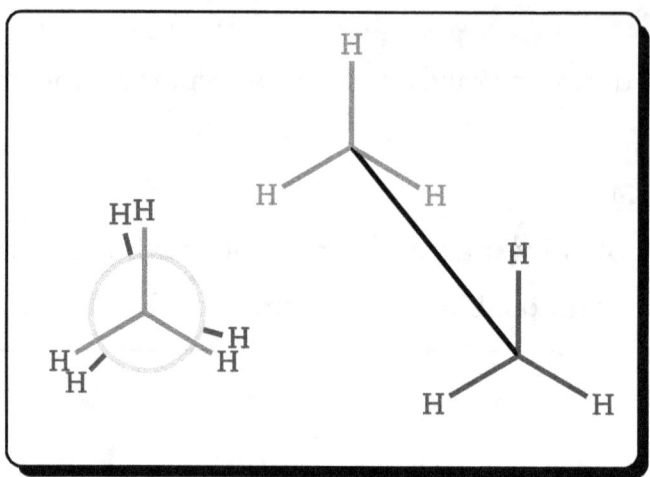

21) RESOLUTION :

The process by which the enantiomers of the racemic forms are separated.

22) ROTAMERS :

Rotamers are extreme conformations.

23) HOMOMERS :

Identical representation of same compound.

STEROISOMERISM

Stereochemistry deals with the study of arrangement of atoms of a molecule in three dimension space.

Stereoisomers are the compounds *that have same molecular formula and mode of attachment between atoms but differ in arrangement of atoms in space.* This phenomenon is called **stereoisomerism**. Stereoisomerism is of three types:

1) Conformational isomerism

2) Optical isomerism

3) Geometrical isomerism.

CONFORMATION ISOMERISM

Groups bonded only by a sigma bond (i.e. by a single bond) can undergo rotation about that bond with respect to each other. That is, these groups are not fixed in a single position, but are relatively free to rotate about the single bond connecting them. The different structures which result by rotation about single bond. Conformational stereoisomers interconvert easily at room temperature through rotation about single bond. A single conformation cannot be physically separated from one another.

METHODS OF REPRESENTING CONFORMATIONS

Three dimensional figures cannot be properly represented on a two dimensional surface like that of paper. Hence various projectional formula have been suggested for drawing conformations. Four of them are discussed below:

❖ **DOTTED LINE WEDGE FORMULA :**

Dotted line wedge formula is a shorthand notation used to simplify three dimensional drawing. In this method the molecule is seen from the side of C – C bond.

1) Dashed lines show the bonds that go backward away from the reader.

2) Thick lines are used to show bonds that come forward, toward the reader.

3) The normal lines represent bonds within the plane of the paper.

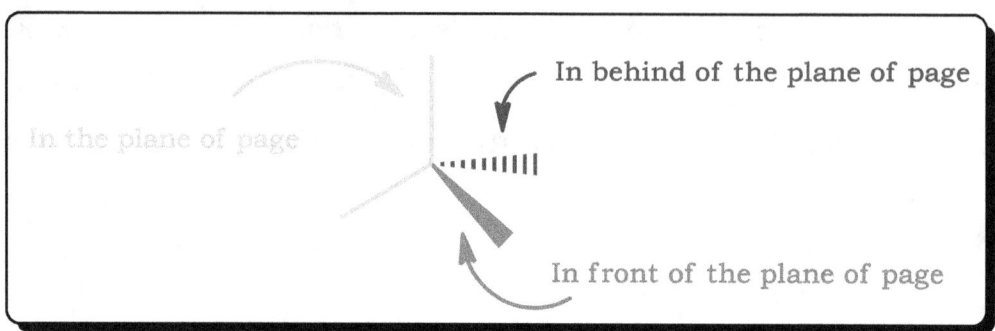

❖ FISCHER PROJECTION:

Fischer projection is a standard way to project three dimensional configuration of each carbon onto a plane surface. The projection looks like a cross with a chiral carbon at the point wedge the lines cross. The Fischer projection can be written in the following way:

If a molecule has a single chiral center e.g. - D-glyceraldehyde, the carbon chain is drawn vertical with the most oxidized atom at the top. Then mentally flatten the structure at each chiral center onto a plane surface. The horizontal lines at the chiral center represent bonds that project forward. The vertical lines at the chiral center actually represent bonds which project in the rear.

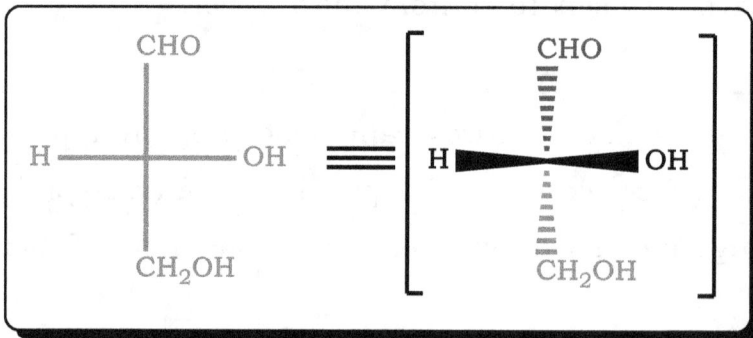

Fischer formula can be used for two or more chiral centers with, conventionally, the longest carbon chain vertical, the bonds to top and bottom atom or groups go back, and all the atoms or groups to one side or the other come forward.

Fischer projection formulae is convenient method for representing acyclic compounds with multiple chiral centers.

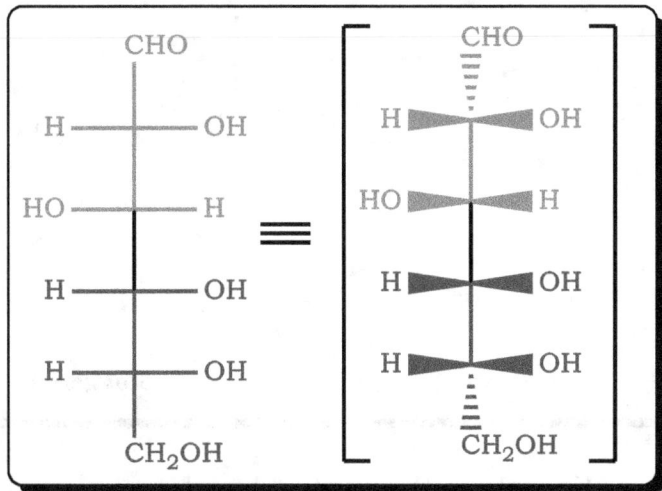

❖ NEWMAN PROJECTION:

Newman's projections are a way of drawing conformation. To draw this formula we have to look C-C bond from the front. The front with its three substituents is denoted by inverted Y. The back carbon is represented by a circle with three bonds pointing out from its periphery.

Three diminsional projection formula

Back carbon atom

Front carbon atom

Newman projection formula

❖ SAW HORSE FORMULA :

Sawhorse formula is the perspective formula which is used to specify a conformation. In sawhorse representation, the bond between C-C atoms is shown by a longer diagonal line. The bonds linking other substituents to these carbons are shown projecting above or below this line.

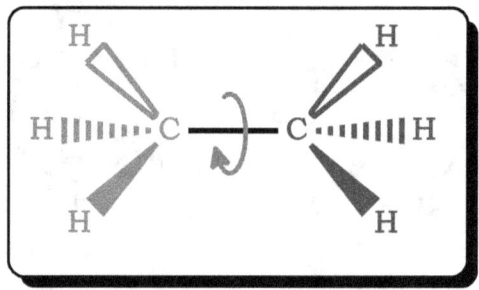

Eclipsed   **Staggered**

Due to free rotation along the central bond, two extreme conformations are possible – the staggered and eclipsed.

CONFORMATIONS OF ETHANE

The two methyl groups in ethane are not fixed in a single position. They are free to rotate about the single bond connecting the two carbon atoms. *The various structures which result by rotation about a single bond are called as* **conformations**.

An infinite number of different conformations could result from rotations of the CH_3 groups about the carbon – carbon sigma bond since the dihedral angle between the hydrogen atoms on the front and back carbon's can have an infinite numbers of values. The two extreme conformations of ethane are

1) Staggered Conformation
2) Eclipsed Conformation

There are innumerable number of conformations in between these two extreme fond called as skew conformations.

1) STAGGERED CONFORMATION:

In the staggered conformation, dihedral angle (0) i.e. the angle between C-H bonds on the front carbon atom and the C-H bonds on the back carbon in the Newman projection, is 60°.

$\theta = 60^0$

Staggered

Stability:

The staggered conformation allows the maximum possible separation of the electron pairs of the six carbon-hydrogen bonds thereby minimizing the repulsive interactions between bonding pairs of electrons and therefore it has lowest energy.

2) ECLIPSED CONFORMATION:

The conformation of ethane with the dihedral angle of 0° is called eclipsed conformation. Newman projection of eclipsed conformation shows the hydrogen atoms on the back carbon to be hidden (eclipsed) by those on the front carbon.

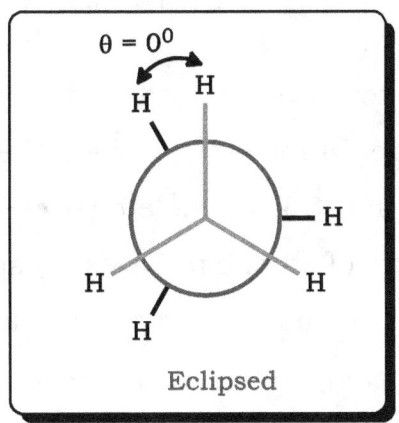

$\theta = 0^0$

Eclipsed

Stability:

In the eclipsed conformation, the electron pairs of the six carbon-hydrogen bonds are closest and therefore eclipsing leads to steric repulsion of hydrogen atoms that are not directly bonded. These non-bonded interactions raises the energy of eclipsed conformation by about 3 kcal/mol. It is highest energy and has the least stability.

3) SKEW CONFORMATION:

Any conformation of ethane that is not precisely staggered nor eclipsed is called as skew Conformation.

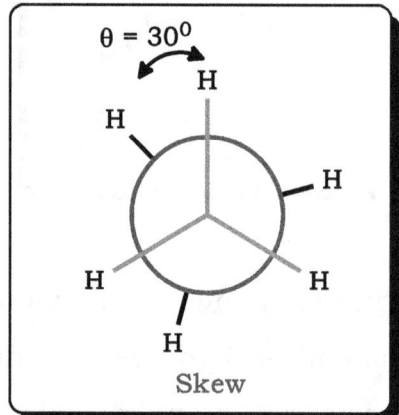

Stability:

In any Skew conformation of ethane, electron pairs of the carbon-hydrogen bonds are not so far as in staggered conformation nor so close as in eclipsed conformation and therefore it is more stable than eclipsed conformation and less stable than staggered conformation.

TORSIONAL ENERGY

When ethane rotate towards eclipsed conformation, its potential energy increases which leads to resistance to rotation. The resistance to twisting is called as torsional strain and 3 kcal/mole energy needed is called torsional energy. The torsional energy of ethane in lowest in staggered conformation. In the eclipsed conformation, the molecule is about 3 kcal/mol higher in energy. This barrier is easily overcome at room temperature and the molecules rotate constantly.

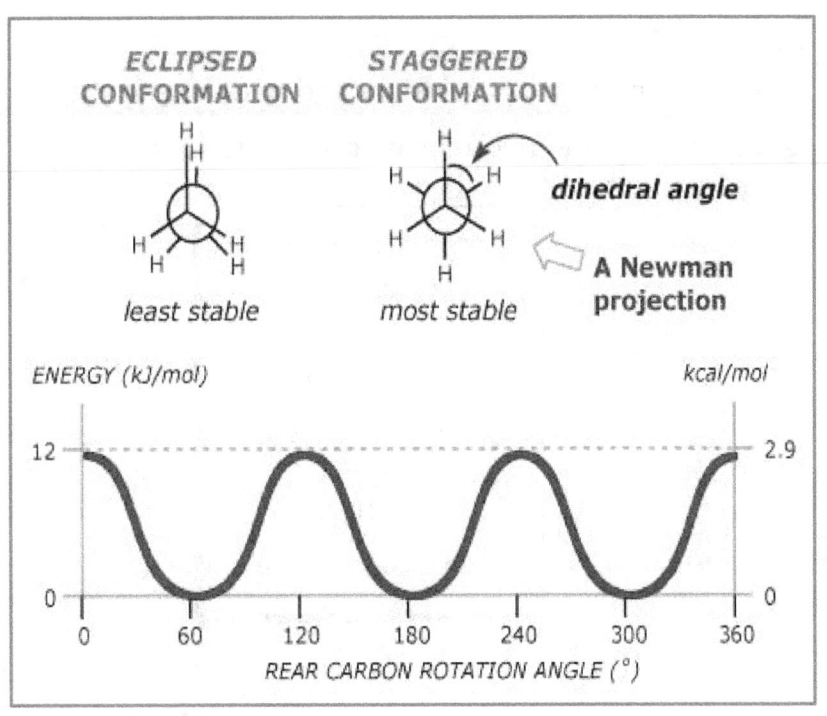

CONFORMATIONS OF BUTANE

Focusing our attention on the middle C-C bond in butane, we see a molecule similar to ethane but with a methyl group replacing one hydrogen on each carbon.

Due to the presence of the two methyl groups, two new point are encountered here

1) There are several staggered and eclipsed conformations.

2) Van der Waals repulsion besides torsional strain plays important role in conformational stabilities.

Various conformations of butane are discussed below:

1) ANTI CONFORMATION:

In anti-conformation there is 180° dihedral angle between the largest groups. e.g. dihedral angle two methyl groups is 180°.

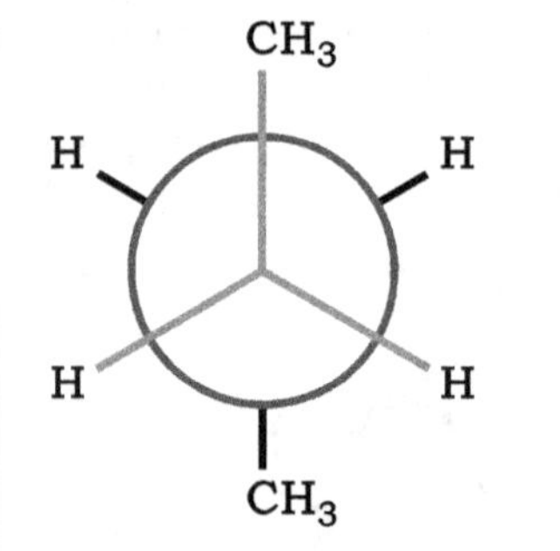

Anti-conformation

Stability:

The anti-conformation does not have torsional strain because *a) the groups are staggered b) two methyl groups are far apart from each other.* Therefore anti-conformation is the most stable.

2) GAUCHE CONFORMATION:

A conformation with a 60° dihedral angle between the largest groups is called gauche conformation. In the gauche conformation of butane, two methyl groups are 60° apart.

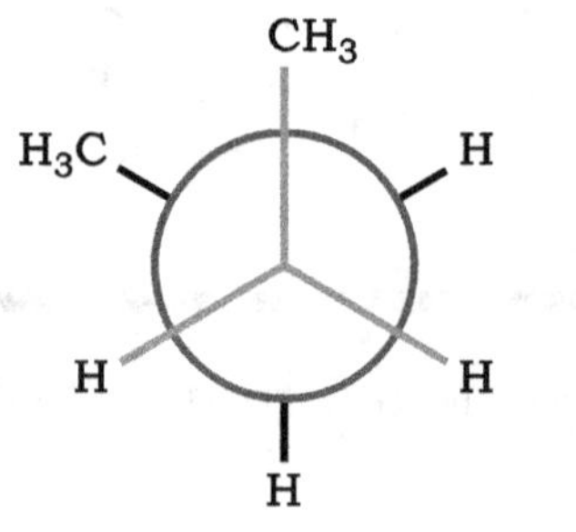

Gauche conformation

Stability:

In the gauche conformation, the methyl groups are close enough to each other that the Van der Waals forces between them are repulsive. This repulsion causes the gauche conformation to have approximately 3.8 kJ/mol energy more than the anti-conformation.

3) PARTLY ECLIPSED CONFORMATION:

Eclipsed conformation of butane in which dihedral angle between two methyl groups is 120° is known as partly eclipsed conformation.

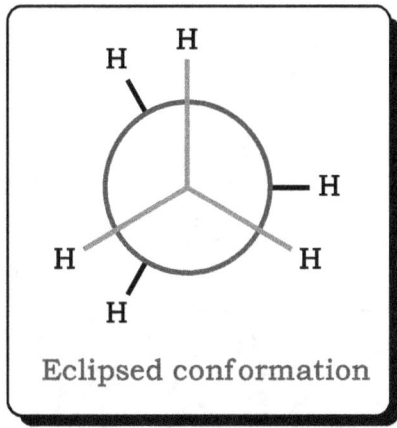

Eclipsed conformation

Stability:

Partly eclipsed conformation has torsional strain. It also has additional van der Waals repulsions arising from the eclipsed methyl groups and hydrogen atoms. These repulsions cause partly eclipsed conformation to have energy 16 kJ/mol than the anti-conformation.

4) FULLY ECLIPSED CONFORMATION:

The conformation in which the methyl groups are pointed in the same direction (dihedral angle = 0°) is called fully eclipsed conformation.

Eclipsed conformation

Stability:

Fully eclipsed conformation has the greatest energy and thus least stability of all because, in the addition to torsional strain, there is added large van der Waal's repulsive forces between the eclipsed methyl groups.

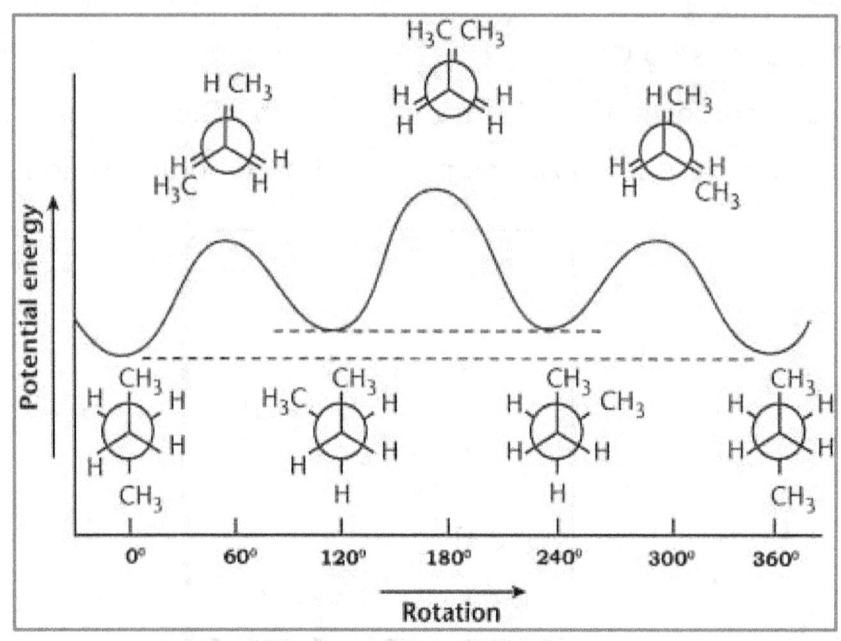

CONFORMATIONS OF CYCLOHEXANE

Existence of an innumerable number of conformers in the case of cyclohexane as happens in the case of ethane or butane is not possible. The number of conformers in the case of cyclohexane are limited due to the presence of ring structure which is rigid.

Various conformations of cyclohexane are discussed below:

1) CHAIR CONFORMATION:

The chair conformation is free from angle strain since each angle is 109° 28'. In chair form all the C-H bonds are in staggered condition and thus it is free of torsional strain. Thus the potential energy of chair form is minimum.

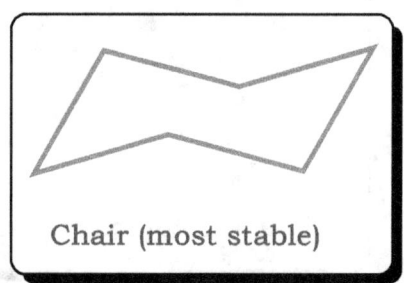

Chair (most stable)

2) BOAT CONFORMATION:

The boat conformation is free from the angle strain since each angle is 109° 28'. Two flagpole hydrogen atoms in the boat conformation lie only 1.83 A° and thus there is steric repulsion between them. Moreover C_2 and C_6 are eclipsing C_3 and C_5 to which there is bond eclipsing strain. The strain energy calculations indicate that the boat conformation is about 6.4 kcal/mol higher in energy than the chair conformation.

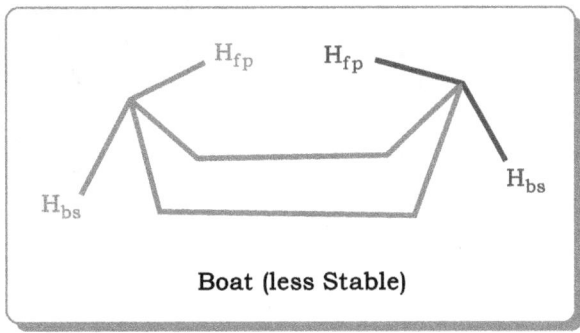

Boat (less Stable)

3) SKEW BOAT CONFORMATION:

When the two flagpole bonds in the boat conformation are moved apart, one gets a twist or skew conformation.

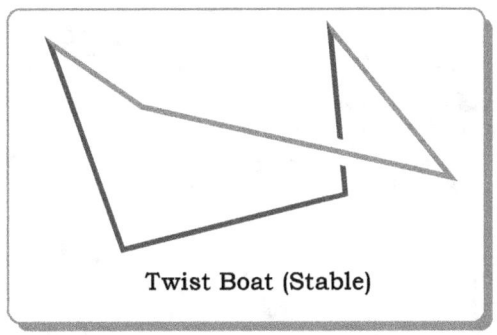

Twist Boat (Stable)

In the skew boat conformation, the flag pole hydrogens are thrown apart. In the first conformer C_2 and C_5 and in the second twist form C_3 and C_6 have gone down. Thus there is less strain in twist conformer than in the boat as there is less of hydrogen eclipsing and flagpole interactions. According to Hendrickson, the twist form contains 1.6 kcal/mol less energy than the boat form. The strain energy calculations indicate that the skew boat conformation is about 5 kcal/mol higher in energy than the chair conformation.

4) HALF CHAIR CONFORMATION:

A transition state conformer between the chair and twist forms is supposed to exist. This is known as half chair conformer. This has a high strain. It has about 11 k cal/mole energy than the chair form and is least stable.

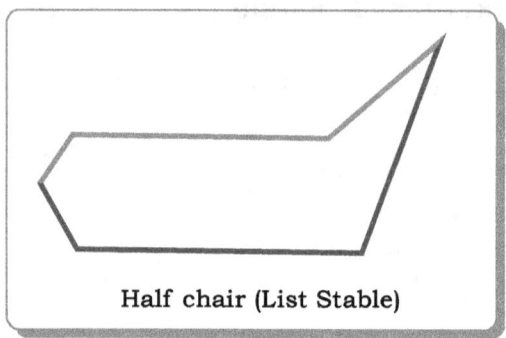

Half chair (List Stable)

We can summaries the relative stabilities of the various conformations of cyclohexane as:

Chair > twist / skew boat > boat > half chair

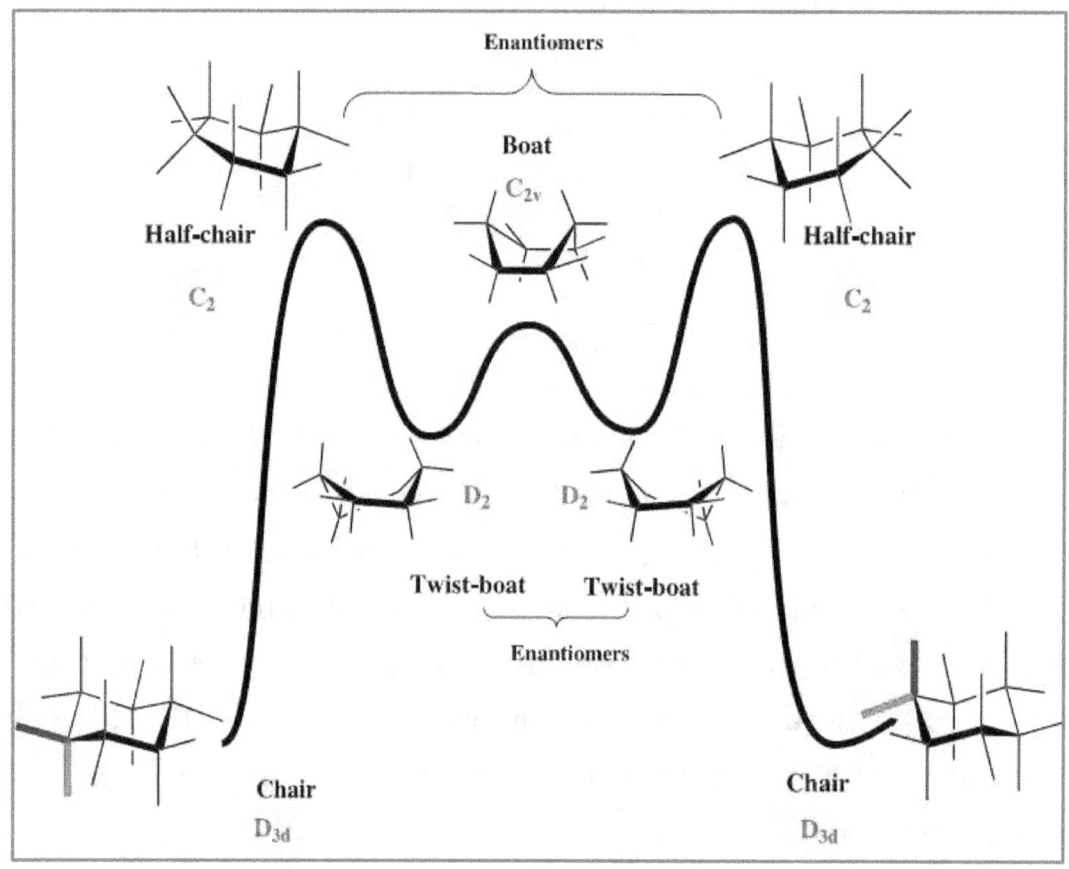

AXIAL AND EQUATORIAL BONDS IN CYCLOHEXANE:

The bonds which are parallel to the three fold axis of symmetry of the chair are known as axial bonds and those which extend outward from the ring are known as equatorial bonds.

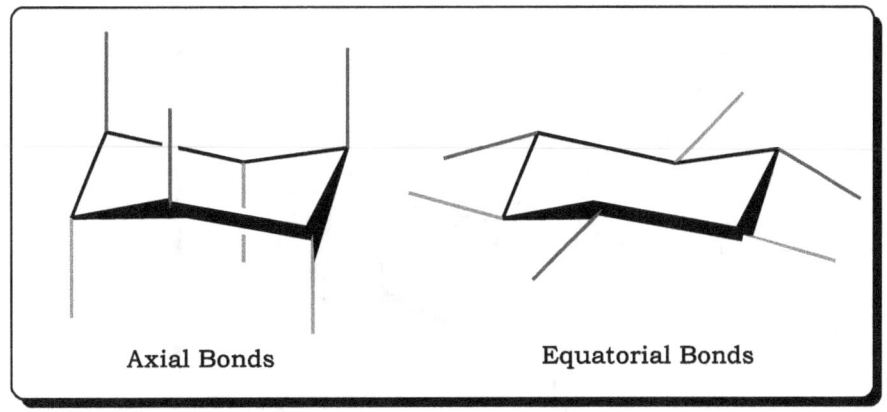

Axial Bonds   Equatorial Bonds

Each carbon atom of cyclohexane has one axial bond and one equatorial bond.

In the case of cyclohexane derivatives when one hydrogen atom is replaced by a larger group or atom then the two chair forms are obtained in case of this mono substituted derivative. The two isomeric chair forms differ in the position of substituent. In one isomer, the substituent is axially located whereas in the other, the substituent is equatorially located. The stabilities of both the forms are different. Let us consider the case of methyl cyclohexane. This molecule can have two isomeric chair forms whose stabilities would be different. The two isomeric chair forms are 1) Axial isomer 2) Equatorial isomer.

1) AXIAL ISOMER:

In the axial conformer the methyl group is located at axial position. In this conformer, the methyl group is so close to the two axial hydrogens on the same side of the molecule (attached to C-3 and C-5 atoms) that the van der Waals forces between them are repulsive. This type of steric strain, because it arises from an interaction between axial groups on carbon atoms that have 1, 3-relation, is called as a 1, 3-diaxial interaction.

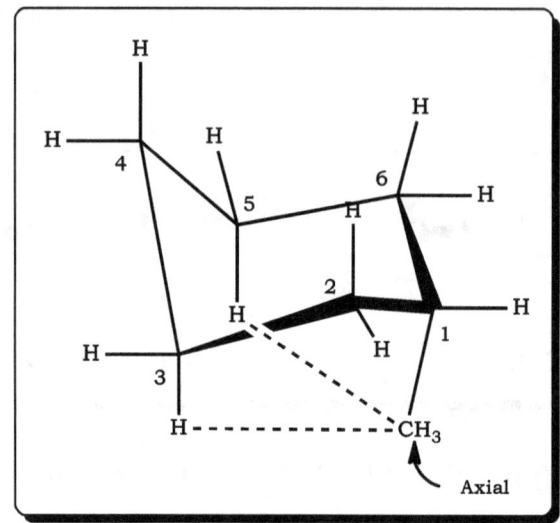

1,3-diaxial methyl-hydrogen interaction is about 0.9 kcal/mole. The strain caused by a 1,3-diaxial interaction in methyl cyclohexane is the same as the strain caused by the close proximity of the hydrogen atoms of methyl groups in the gauche form of butane. These gauche interactions in the gauche butane causes gauche butane to be less stable than anti-butane by 3.8 kJ/mol.

Gauche-Buatne
(3.8 kJ/mol steric strain)

Axial methyl cyclohexane
(two gauche interactions = 7.6 kJ/mol steric strain)

2) EQUATORIAL ISOMER:

In the equatorial isomer, the methyl group is placed positions. In the equatorial conformer the methyl group extends into space away from the rest of molecule because of which its hydrogen atoms are far away. It is free from diaxial interactions because equatorial methyl group is anti to C-3 and C-5.

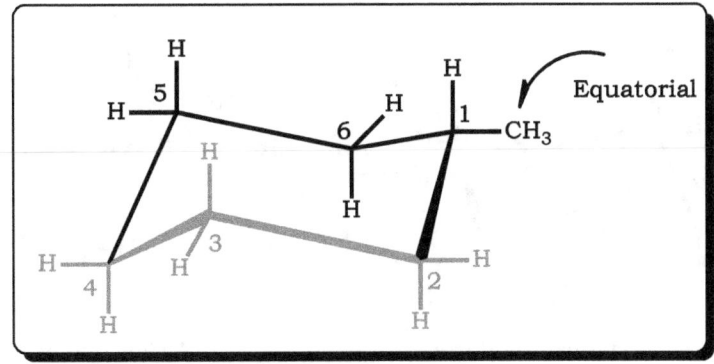

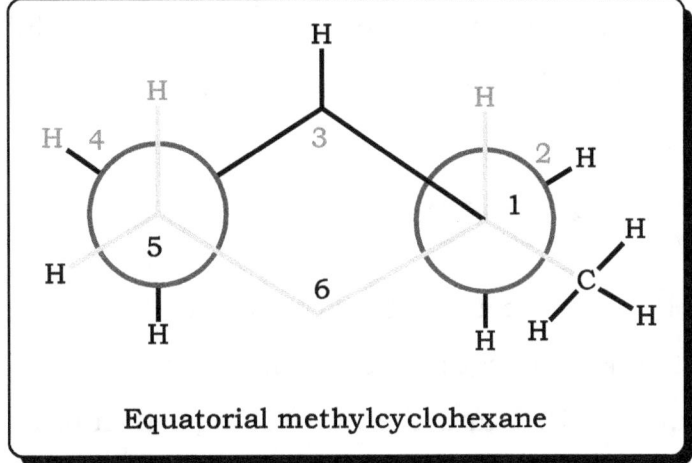

Equatorial methylcyclohexane

Stability:

Equatorial conformer is more stable than axial conformer.

MAGNITUDE OF 1, 3-DIAXIAL INTERACTION

The magnitude of the 1,3-axial interactions varies with different substituents. The energy difference between the axial and equatorial conformers can be larger or smaller depending on the substituent on the ring.

CONFORMATIONAL FREE ENERGY:

The energy difference between conformers is known as conformational free energy. The important aspect of conformational analysis is that the two diastereomeric chair forms are not of equal free energy and therefore are differently populated. In other words we can say that the different mono-substituted cyclohexane derivatives display different conformational preferences due to difference in energies of axial and equatorial conformers.

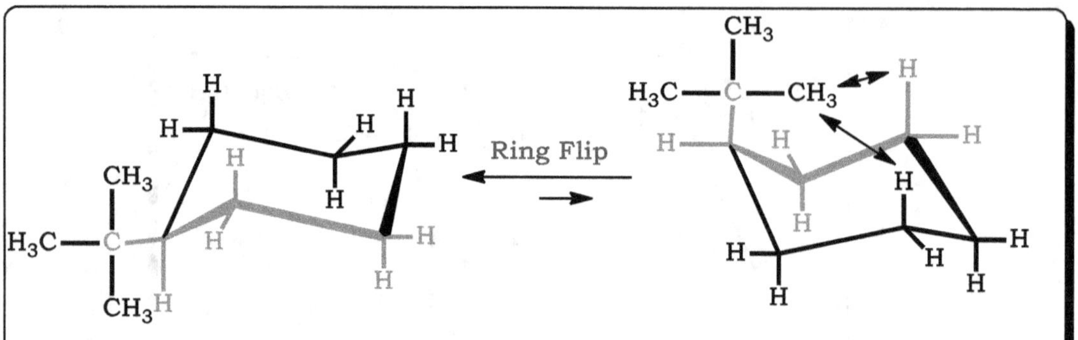

Equatorial tert-butylcyclohexane   Axial tert-butylcyclohexane

There is a direct relationship between difference in energy, called the free energy ($\Delta G°$) and the equilibrium constant (K_{eq}) associated with a given equilibrium in solution

$$\Delta G° = \textit{difference in free energy} = RT \ln k_{eq}$$

Where, R = gas constant (0.00199 kcal/mole) and T = absolute temperature at which the equilibrium is measured.

The product of R, T and the natural logarithm of k_{eq} gives $\Delta G°$ i.e. the free energy difference between the two conformers in kcal/mol. $\Delta G°$ is usually negative is the difference of free energy between the equatorial and axial conformer and $\Delta G°$ is known as conformational free energy of the substituent.

For substituted cyclohexane, it is conventional to specify the value of -$\Delta G°$ for the equilibrium:

Axial ⇌ Equatorial

$\Delta G°$ with be negative when the equatorial conformation is more stable than the axial. The value of $\Delta G°$ is positive for the case of substituent groups which favour the equatorial position. The larger the $\Delta G°$, the greater is the preference for the equatorial position.

CONFORMATIONAL FREE ENERGIES OF SUBSTITUENT GROUPS:

Conformational free energy values for many substituent groups on cyclohexane ring are determined by NMR spectroscopy. Conformational free energy values are measured at low temperatures. It is believed that these values do not vary much at room temperature.

Conformational free energies (-ΔG°) for substituent groups

Substituent	-ΔG° (kcal/mole)	Substituent	-ΔG° (kcal/mole)
-F	0.24-0.28	$-C_6H_5$	2.9
-Cl	0.53	-OH (Aprotic solvents)	0.52
-Br	0.48	-OH (Protic solvents)	0.87
-I	0.47	$-OCH_3$	0.60
$-CH_3$	1.8	$-NO_2$	1.16
$-CH_2CH_3$	1.8	-CN	0.15-0.25
$-CH(CH_3)_2$	2.1	$-O_2CCH_3$	0.71
$-C(CH_3)_3$	>4.5	$-CO_2H$	1.35
$-CH=CH_2$	1.7	$-CO_2C_2H_5$	1.1-1.2
-HgBr	0		

Relationship between free-energy difference and isomer percentage for isomers at equilibrium at 25°C

Free-Energy Difference, ΔG° (kJ/mol)	More Stable Isomer (%)	Less Stable Isomer (%)
0	50	50
1.7	67	33
2.7	75	25
3.4	80	20
4.0	83	17
5.9	91	9
7.5	95	5
11	99	1
17	99.9	0.1
23	99.99	0.01

DISUBSTITUTED CYCLOHEXANE

The presence of two substituents on the ring of a cyclohexane allows for the possibility of cis-trans isomerism. Geometrical isomerism in various di-substituted cyclohexane is discussed below.

1,2 – DISUBSTITUTED CYCLOHEXANE :

The planner representation of cis and trans isomers of 1,2 – disubstituted cyclohexane is as follows

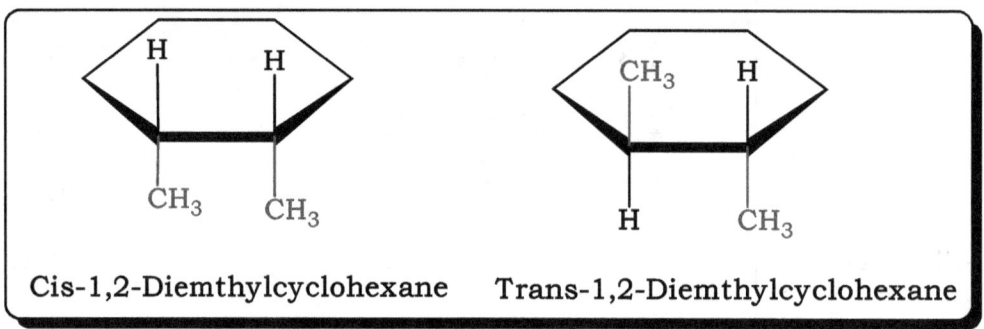

Cis-1,2-Diemthylcyclohexane   Trans-1,2-Diemthylcyclohexane

❖ CIS FORM:

The cis isomer has two identical *equatorial and axial and axial and equatorial* conformations. Cis conformation has three butane gauche interactions. Consider cis-1, 2-dimethyl cyclohexane, the two conformations are shown below.

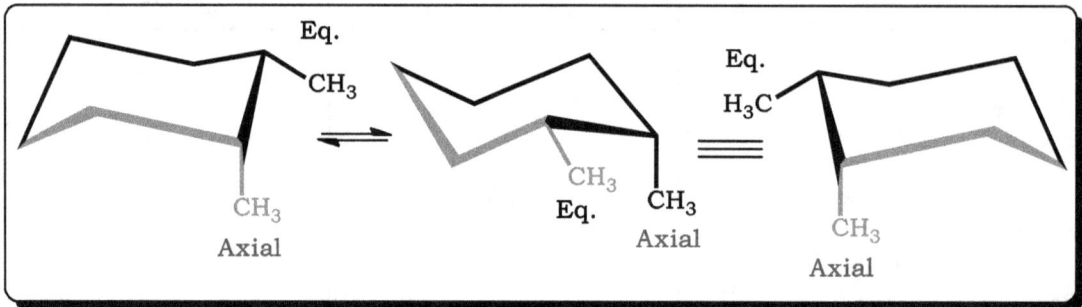

❖ TRANS FORM:

There are two possible chair conformations of trans-1, 2- disubstituted cyclohexane. In one conformation, both the groups are axial; in the other both are equatorial. The two chair conformations of 1, 2- dimethyl cyclohexane are

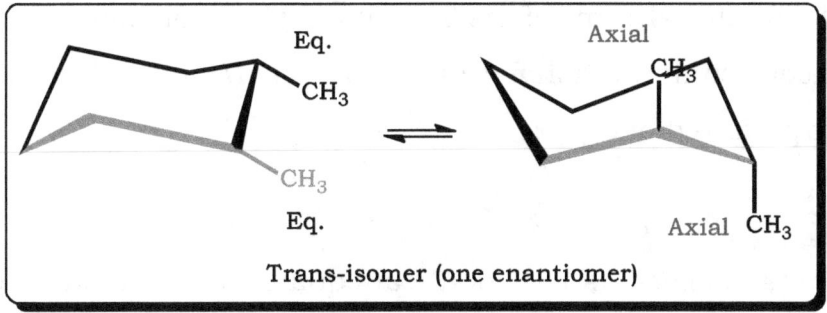

Axial

Eq.

CH₃

Axial CH₃

CH₃

Eq.

Axial CH₃

Trans-isomer (one enantiomer)

The diaxial form of trans-isomer in the case of 1,2-dimethyl cyclohexane has four butane gauche interactions whereas the di-equatorial form has only one, that between the methyl groups. Thus, in the case of 1,2-dimethyl cyclohexane, the di-equatorial trans isomer is more stable than the cis isomer by about 1.8 kcal/mole.

1,3 – DISUBSTITUTED CYCLOHEXANE

1,3-disubstituted cyclohexane exist in diastereomeric cis and trans forms whose planar representation is as follows

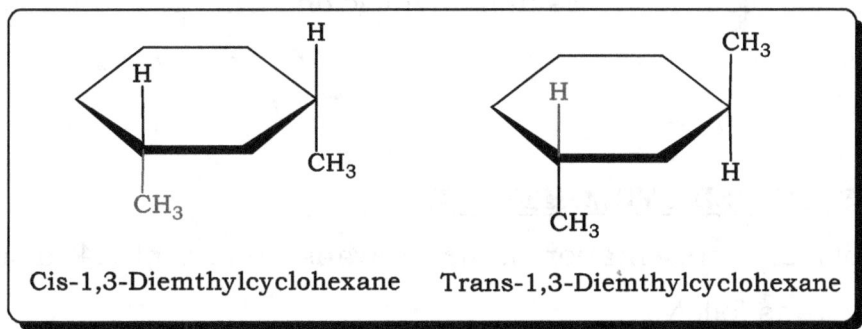

Cis-1,3-Diemthylcyclohexane   Trans-1,3-Diemthylcyclohexane

❖ CIS FORM:

There are two possible conformations of cis form. In one conformation, both the groups are axial whereas in the other, both the equatorial. The two possible conformations of cis-1,3- dimethyl cyclohexane are given below.

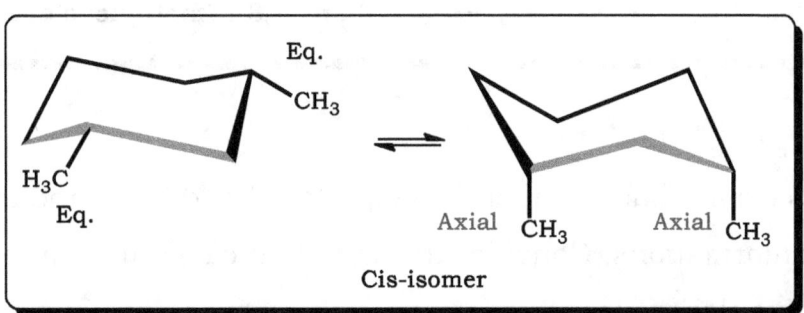

Cis-isomer

The di-equatorial form of cis-1,3-dimethyl cyclohexane is more stable than the diaxial conformation by about 5.4 kcal/mole. Thus di-equatorial form is most preferred one.

❖ **TRANS FORM:**

The trans isomer has two identical equatorial and axial and axial and equatorial conformations. Consider trans-1,3- dimethyl cyclohexane with identical conformations.

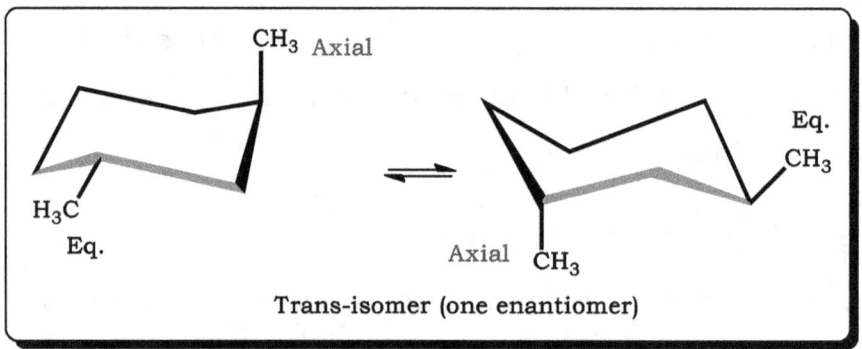

Trans-isomer (one enantiomer)

The trans isomer of 1,3- dimethyl cyclohexane has to butane gauche interactions. *Thus the cis isomer is more stable by about 1.8 kcal/mole than the trans isomer.*

1.4 DISUBSTITUTED CYCLOHEXANE:

The planar representation of the cis-trans isomers of 1,4- disubstituted cyclohexane is as follows

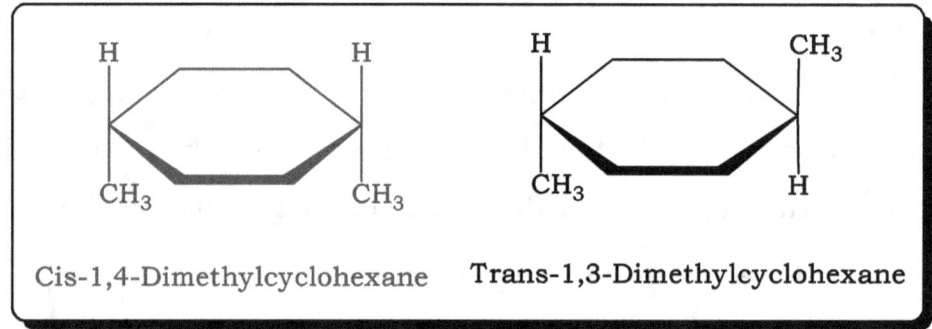

Cis-1,4-Dimethylcyclohexane   Trans-1,3-Dimethylcyclohexane

❖ **CIS FORM :**

The cis isomer has two identical equatorial and axial and axial and equatorial conformations. Consider the two identical conformations of cis-1,4- dimethyl cyclohexane.

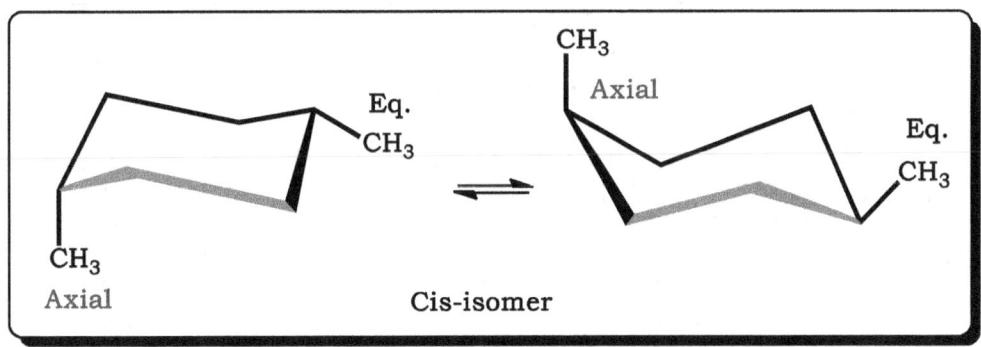

TRANS FORM:

There are two possible chair conformations of trans for of 1,4 di-substituted cyclohexane. In one conformation both the groups are axial; in other both are equatorial. Consider the two conformations of trans-1,4-dimethyl cyclohexane.

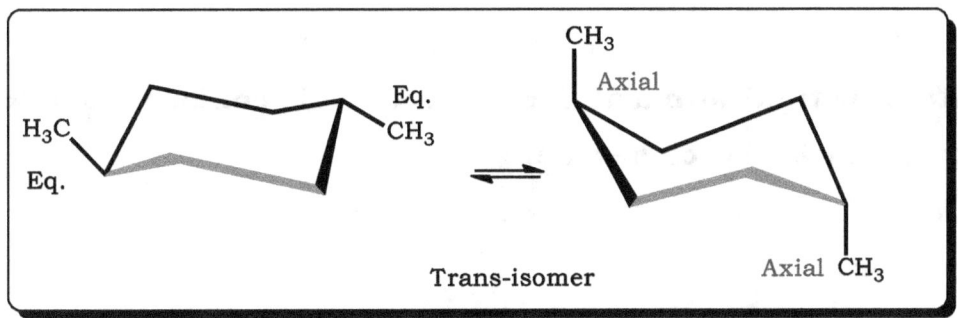

The di-equatorial conformation is more stable than diaxial conformation and it represents the structure of at least 99% of the molecules equilibrium. The diaxial conformation has four gauche interactions.

Relationship between free-energy difference and isomer percentage for isomers at equilibrium at 25°C.

Isomer	Conformation	No. of gauche interactions	Interaction kcal/mole
cis-1,2	Equatorial, Axial ↕ Axial, Equatorial	3	2.7
trans-1,2	Equatorial, Equatorial ↕ Axial, Axial	1	0.9
		4	3.6

cis-1,3	Axial, Axial ↓↑ Equatorial, Equatorial	0	0
trans-1,3	Equatorial, Axial ↓↑ Axial, Equatorial	2	1.8
		2	1.8
cis-1,4	Equatorial, Axial ↓↑ Axial, Equatorial	2	1.8
trans-1,4	Equatorial, Equatorial ↓↑ Axial, Axial	0	0
		2	1.8

PROBLEM:

a) Write structural formulas for the two chair conformations of cis-1-isopropyl-4-methyl cyclohexane.

b) Are these two conformations equivalent?

c) If not, which would be more stable? d) Which would be the preferred conformation at equilibrium?

Solution:

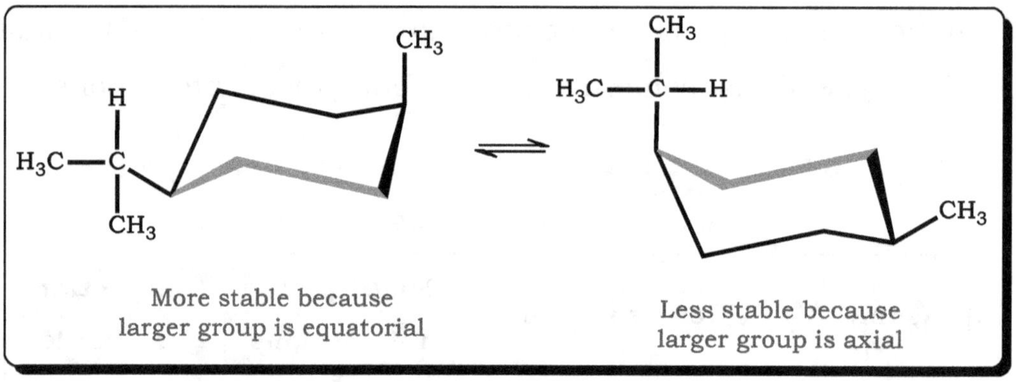

More stable because larger group is equatorial

Less stable because larger group is axial

PROBLEM:

a) Write the two conformations of cis-1,2-dimethylcyclohexane.

b) Would these two conformations have equal potential energy?

c) What about the two conformations of cis-1-terta-buty1-2-methylcyclohexane?

d) Would the two conformations of trans-1,2dimethylcyclohexane have the same potential energy?

Answer (a):

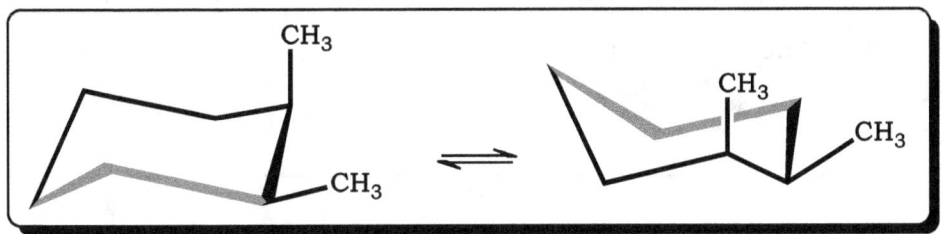

Answer (b): Yes

Answer (c):

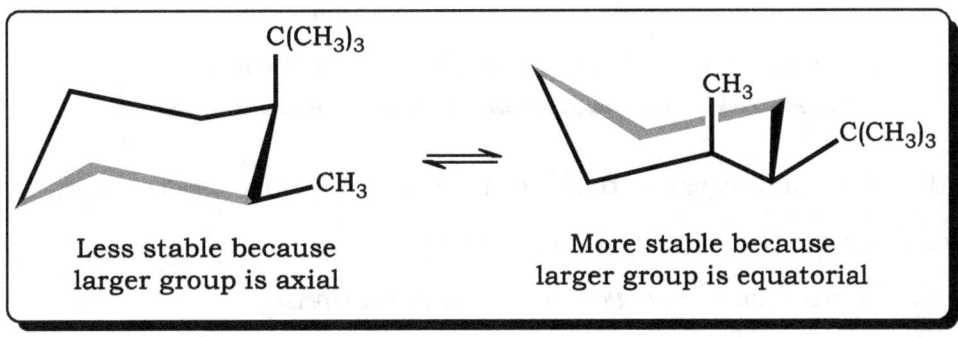

Less stable because larger group is axial

More stable because larger group is equatorial

Answer (d):

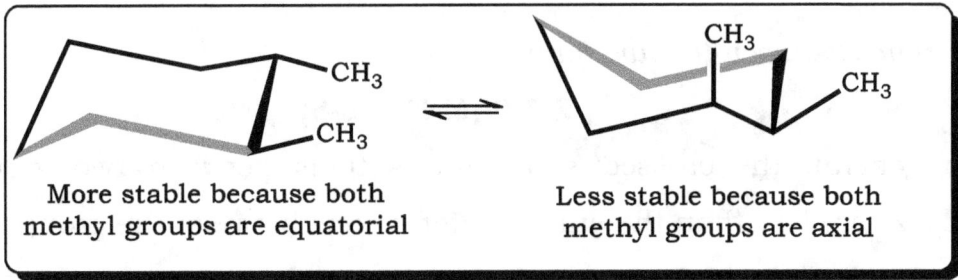

More stable because both methyl groups are equatorial

Less stable because both methyl groups are axial

CONFORMATIONAL EFFECTS ON STABILITY

The free energies of acyclic diasteroisomers usually differ. Generally meso forms are more stable than d *l* pairs. This is illustrated by considering meso isomer and its active diastereoisomer in their most stable conformations.

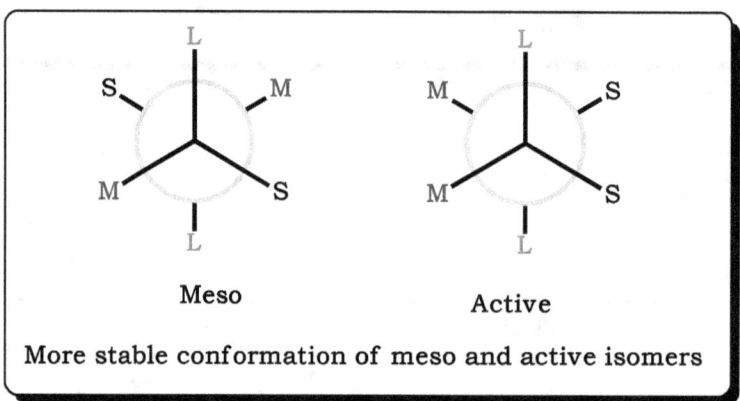

Meso

Active

More stable conformation of meso and active isomers

L denotes the largest substituent, M the medium sized substituent, and S the small substituent in the two isomers.

The gauche interactions observed in the meso form are

$$2L\text{-}M + 2L\text{-}S + 2M\text{-}S + 2L\text{-}M + 2L\text{-}S$$

The gauche interactions observed in the active form are

$$2M\text{-}S + 2L\text{-}S + M\text{-}M + S\text{-}S$$

The difference between the two form is

$$2M\text{-}S\text{-}(M\text{-}M + S\text{-}S)$$

In general, the crossed steric interactions between two groups of unequal size are less than the sum of interactions between the groups of like size i.e. **(M-M+S-S)>2M-S**, provided that interactions are purely steric in origin.

Thus it is follows from the above discussion that meso isomer is more stable than the active isomer.

GEOMETRICAL ISOMERISM

Geometrical isomerism is another type of stereoisomerism arising out of different spatial arrangement of groups attached to double bonds or rings in which stereoisomers are not readily interconvertible.

GEOMETRICAL ISOMERS:

Geometrical isomers are *stereoisomers which differ in spatial arrangement of atoms or groups attached to double bonds or rings and this phenomenon is known as* **geometrical isomerism**.

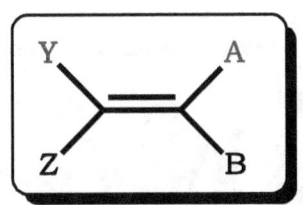

RESTRICTED ROTATION AND THE DOUBLE BOND:

Substituents attached to the C = C can't rotate freely since there is large energy barrier to rotation associated with the groups joined by a double bond.

REASON FOR HINDERED ROTATION:

The C = C double bond consist of *Π* bond and σ bond; it is difficult to rotate the substituents 180°, since the *Π* bond must be broken, a reaction which requires about 264 kJ/mol of energy. Such a rotation will seldom happen at room temperature. The inability of an olefinic double bond to rotate at room temperature is called hindered rotation.

GEOMETRIC ISOMERISM IN OLEFINS:

There is hindered rotation about any carbon-carbon double bond but not all show geometric isomerism. Geometric isomerism is only observed when there is a certain relationship among the groups attaché to the doubly bonded carbons. The requirement for geometric isomerism is shown in following olefin.

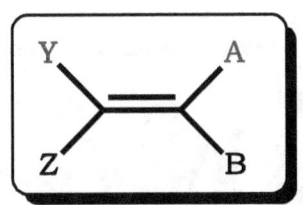

The requirement for geometric isomerism is A and B must be different groups, as must Y and Z; However, either A or B can be same as Y or Z.

Thus, on this basis, we find that propylene, 1-butene and isobutylene do not show isomerism.

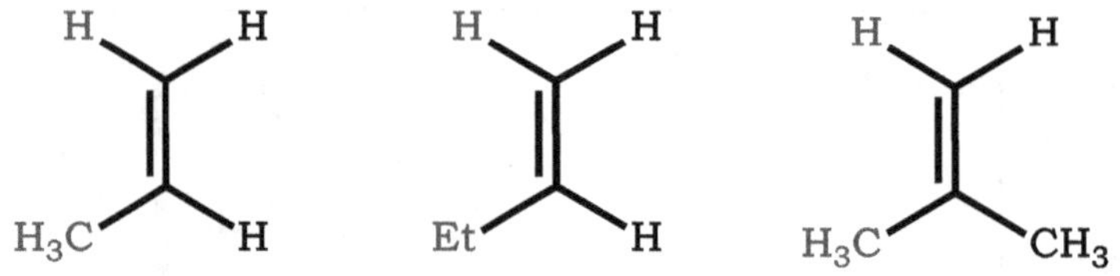

Propylene	1-Butene	Isobutylene

No geometric isomerism

Geometric isomerism can't exist if either carbon carries two identical groups.

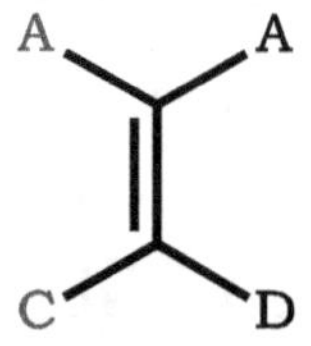

No isomerism

CIS-TRANS ISOMERS:

The prefixes *cis* and *trans* work well to specify the geometric isomers.

❖ **CIS ISOMERS:**

The geometric isomer in which similar groups are present on the same side of the double bond is referred as *cis* isomer.

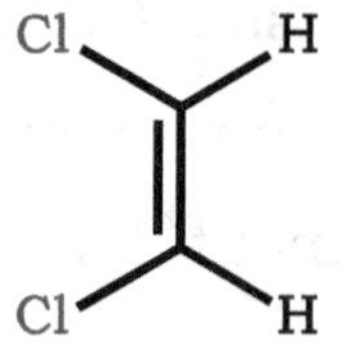

Cis-1,2-Dichloroethnane

❖ TRANS ISOMERS :

The geometric isomer in which similar groups are present on the opposite side of the double bond is referred as *trans isomer.*

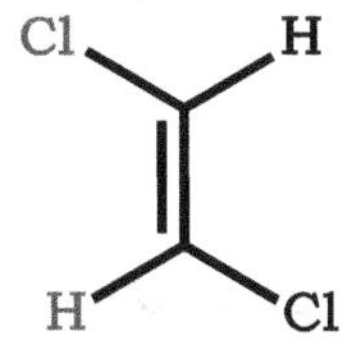

Trans-1,2-Dichloroethnane

DRAWBACKS OF CIS-TRANS NOMENCLATURE:

Cis-trans nomenclature fails to specify the configurations of following compounds.

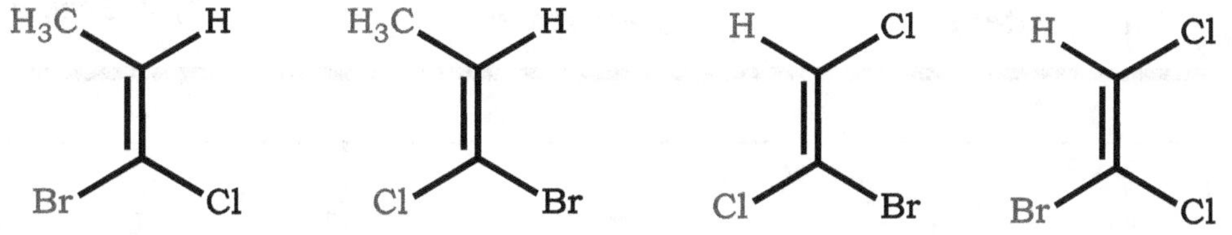

E-Z SYSTEM OF NOMENCLATURE

E-Z system of nomenclature for geometric isomers have been developed after **Cahn-Ingold-Prelog** convention for chiral carbon atoms.

In order to assign E-Z nomenclature to geometric isomer:

❖ The two groups attached to each carbon of the double bond are arranged in order of priority.

❖ If the two groups of highest priority are together on the same side of the double bond then the configuration is called as **Z isomer.**

❖ If the two groups of highest priority are on the opposite side of the double bond then the configuration is called as **E isomer.**

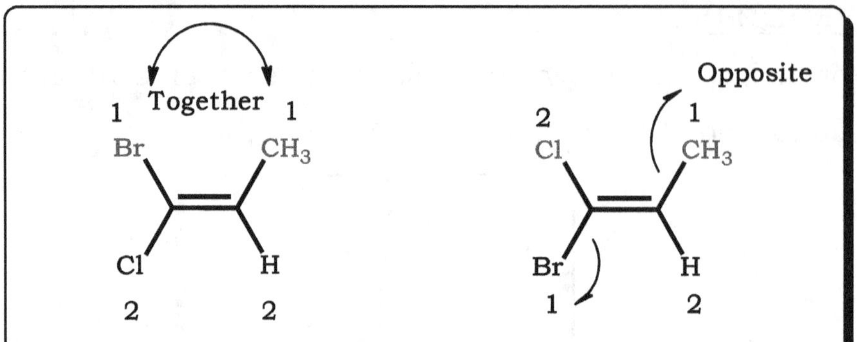

(Z)-1-bromo-1-chloropropane   (E)-1-bromo-1-chloropropane

CH₃ > H   Cl > F

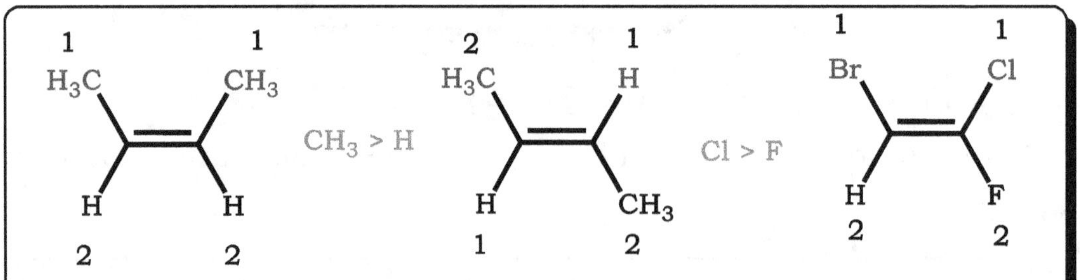

(Z)-2- Butene   (E)-2-Butene   (Z)-2-Butene

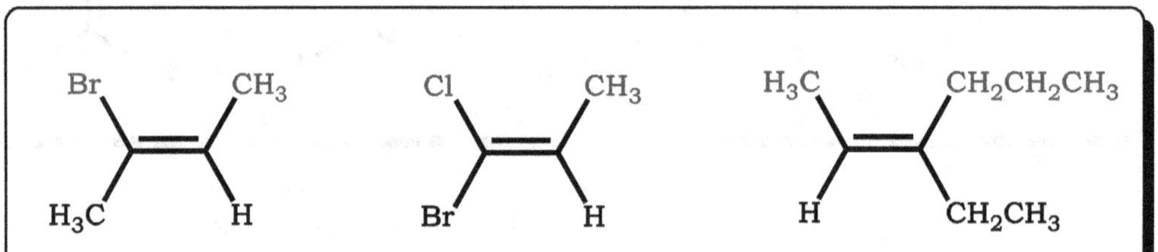

(I)   (II)   (III)

When an alkene has more than one double bond, the stereochemistry about each double bond can be specified with E and Z nomenclature.

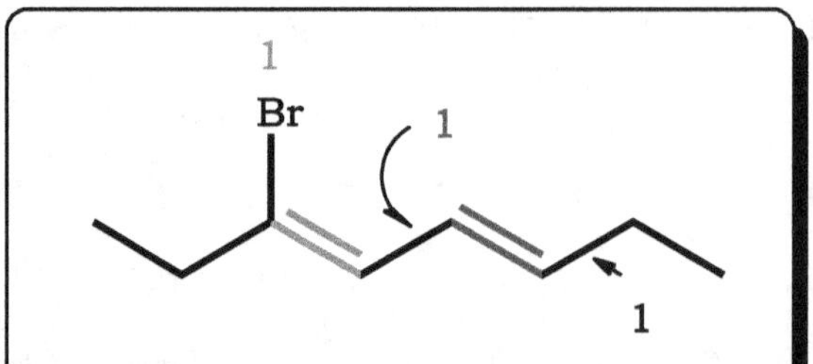

(3-bromo-(3Z, 5E)-octadiene)

The E-Z nomenclature can also be used to designate cyclic compounds. When the two higher priority groups are on the same side of the ring then the compound is called as **Z isomer** and when these groups are on opposite side of the ring, the compound is called as *E isomer*.

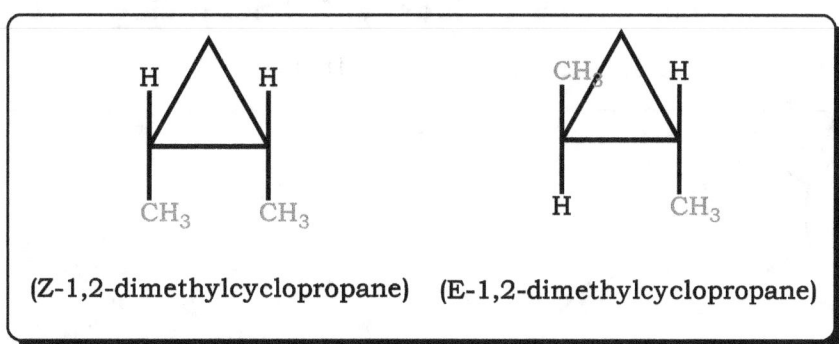

(Z-1,2-dimethylcyclopropane)   (E-1,2-dimethylcyclopropane)

The Z or E isomers are not necessarily always the ones which would be called *cis or trans* isomers respectively under the old nomenclature because E and Z nomenclature depends on the priorities of the groups attached to the unsaturated carbon atoms.

Cl > H

Br > Cl

(E)-1-bromo-1,2-dichloroethane

(Cis)-1-bromo-1,2-dichloroethane

(Z)-1-bromo-1,2-dichloroethane

(Trans)-1-bromo-1,2-dichloroethane

OLEFINS AND CHIRALITY:

The compounds which show cis-trans isomerism with one double bond are not chiral because the four groups are in one plane.

When the compound contains odd number of cumulative double bond (three, five etc.) then orbital overlap causes the four groups to occupy one plane and hence cis-trans isomerism is observed in such compounds.

When the compound contains even number of cumulative double bonds and when the both sides are dissymmetric then optical activity is possible in such compounds.

MULTIPLE DOUBLE BONDS:

If a molecule has more than one double bond, each substituted properly so as to give geometrical isomerism, then ***the number of possible geometric*** *isomers of it will be* 2^n. Thus four geometric isomers should exist if there are two such double bonds.

The four isomers of **5-cyclohexyl-2,4-pentadien-1-ol** are designated as trans-trans, cis-cis, trans-cis and cis-trans.

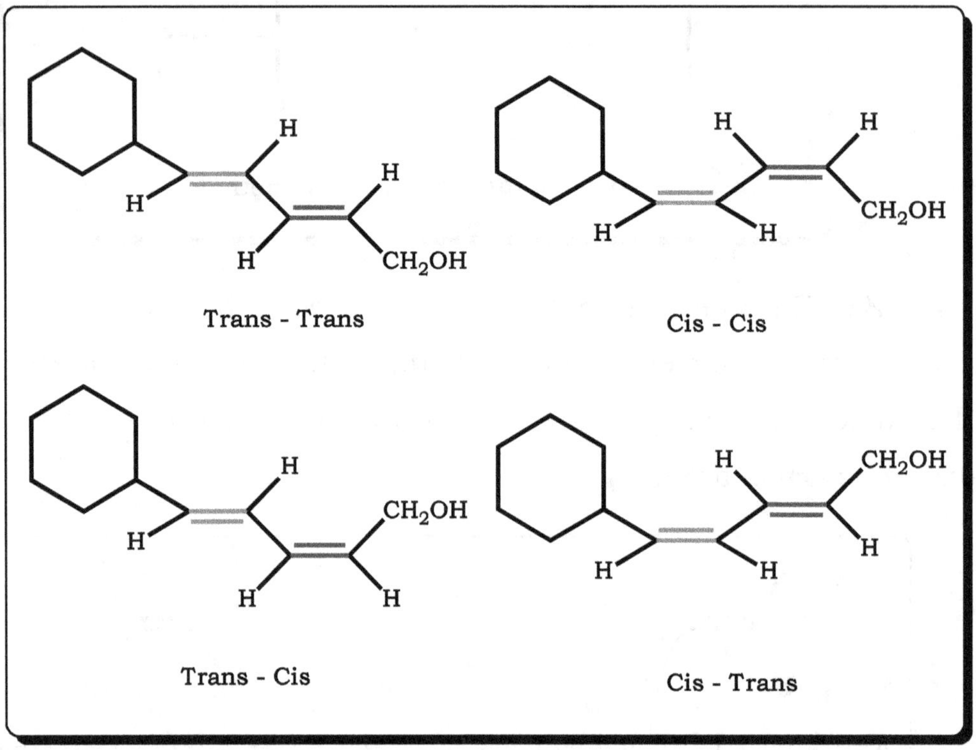

Trans - Trans

Cis - Cis

Trans - Cis

Cis - Trans

Vitamin A has five double bonds thus the total number of possible isomers is $2^5 = 32$.

PROPERTIES OF GEOMETRICAL ISOMERS:

A pair of geometric isomers can be referred as diastereomers. Thus as far as chemical and physical properties are concerned, geometric isomers show the same relationship to each other as do the other diastereomers.

❖ **Chemical Properties:**

The chemical properties of geometrical isomers are not identical, however, since their structures are neither identical nor mirror images; they react with the same reagents, but at different rates. (Under certain conditions-especially in biological systems-geometrical isomers can vary widely in their chemical behavior).

❖ **Physical Properties:**

Geometrical isomers have different physical properties such as melting point, boiling point, refractive indices, solubility, densities etc.

They can be distinguished from each other on the basis of their physical properties. On the basis of the differences in physical properties they can be separated.

INTERCONVERSION OF GEOMETRICAL ISOMERS:

The most straight forward way of interconverting geometrical isomers is by heating. The cis and trans isomers can be interconverted at higher temperatures or by irradiation with light of suitable wavelength.

The interconversion of isomers involves the breaking of the Π bond of the carbon-carbon double bond followed by rotation about the carbon-carbon σ bond and subsequent reformation of a new Π bond.

OPTICAL ACTIVITY

ORDINARY AND PLANE POLARIZED LIGHT:

The nature of light is such that no purely verbal description can adequately represent all of its properties. However one of the oldest and most successful attempts at a description of this phenomenon treats light as a form of energy which is transmitted in waves.

When an ordinary light is passed through a Nicol prism, it is converted into plane polarized light. Plane polarized light can be defined as the light whose waves vibrate in one direction (plane). When ordinary light from a source with an infinite number of planes is passed through a Nicol prism, only a single plane is allowed to emerge.

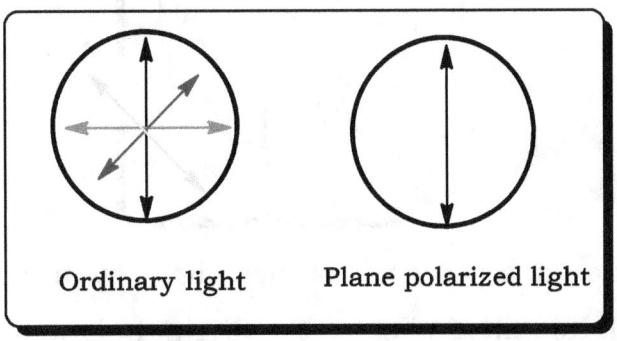

Ordinary light   Plane polarized light

OPTICAL ACTIVITY:

Certain substances have ability to rotate the plane of polarized light are called as optically active compounds and this phenomenon of rotating the plane of polarized light is called *optically activity*. If the substance does not rotate the plane of polarized light, it is considered to be optically inactive.

DEXTRO AND LAEVO ROTATORY SUBSTANCES:

Optically active compound may rotate the plane of polarized light to the right or to the left. The substances which rotate the plane of plane polarized light to the right are called as *dextrorotatory* and are designated as d or (+). The substances which rotate the plane of plane polarized light to the left are called as *laevorotatory* and are designated as *1* (-).

SPECIFIC ROTATION:

The presence of optical activity, its direction and extent of rotation is measured by an instrument called polarimeter. The optical activity of compound is reported as its specific rotation.

OPTICAL ROTATION:

If polarized light is allowed to pass through a solution of an optically active compound then the single plane of polarized light will be rotated. The rotated light is then allowed to pass through a second prism. This prism is rotated until the plane of light is again vertical. The angle that the second prism must be moved to bring the light back to vertical is called as *optical rotation (a).*

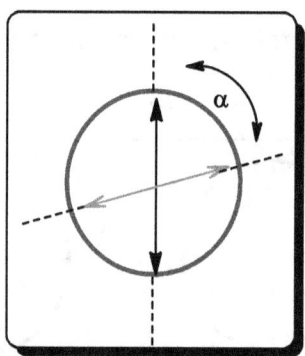

Diagram: Rotation of a plane of polarized light by an optically active organic molecule. The dotted line indicates the original plane of light and *a* degree of rotation from original plane.

Optical rotation is a function of concentration, sample thickness, temperature, wavelength of polarized light etc.

SPECIFIC ROTATION [α]:

Specific rotation is the number of degrees of rotation of the plane polarized light to the wavelength of the sodium D line (5890 A°) when passed through a solution of concentration 1 g/ml in a 1-decimeter tube. Optical rotation is usually recorded in terms of specific rotation. The equation to determine specific rotation [α] is

$$[\alpha] = \frac{\alpha}{C\,l}$$

Where [α] = specific rotation, α = observed rotation, C = the concentration of the solution in g/ml of solution, l = the length of the tube in decimeters (1 dm = 10 cm).

The specific rotation depends on the temperature and the wavelength of light that is employed and hence specific rotations are reported so as to incorporate these quantities. A specific rotation might be given as follow:

$$[\alpha]_0^{25} = +3.12^0$$

This means that the D line of a sodium lamp was used for the light, that a temperature of 25°C maintained and that a sample containing 1 g/ml of the optically active substance, in a 1 dm tube, produced a rotation of 3.12° in a clockwise direction.

The specific rotation is considered as another physical constant like melting point, boiling point, or density.

Problem: The concentration of cholesterol dissolved in CHCI₃ is 6.15 g per 100 ml. of solution. A) A portion of this solution in a 5 cm polarimeter tube causes an observed rotation of – 1.2°. Calculate the specific rotation of cholesterol. B) Predict the observed rotation if the same solution were placed in a 10 cm. tube.

Solution: A] -39°   B] -2.4°

Problem:

An aqueous solution of pure compound of concentration 0.10 g/ml had observed rotation -30° in a 1.0-dm tube at 589.6 nm and 25°C. Determine the specific rotation.

Solution: -300°

CAUSE OF OPTICAL ACTIVITY (Theory of Van't Hoff and Le Bel):

By 1874, over a dozen examples of optically active organic molecule were known. In every case at least one carbon in the molecule had four different groups attached to it.

Van't Haff and Le Bel related the phenomenon of optical rotation to the presence of asymmetrically substituted carbon atoms (chiral carbon atoms) in the molecules.

CHIRAL CARBON:

If the carbon atom is attached to four different groups or atoms, it is called **chiral carbon.**

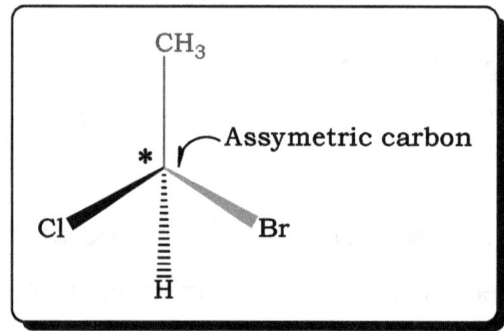

In the case of organic compounds, the presence of chiral carbon is most probable cause of optical activity. They also realized that there could exist optically active compounds having no asymmetric atoms. They also observed that many compounds were optically inactive though they contained two or more asymmetric carbon atoms.

Optically active though doesn't contains asymmetric carbons atoms

Optically inactive though contains two symmetric carbons atoms

Thus the concept of asymmetric carbon atoms could not explain satisfactorily the cause of optical activity.

The theory of Van't Hoff and Le Bel States that, for a molecule or a crystal to be optically active, it's mirror image must be non-superimposable. Whereas a molecule with superimposable mirror image is optically inactive.

NON SUPERIMPOSABLE MIRROR IMAGE:

An object or molecule or crystal can be superimposable on its mirror image when it has any one of the following elements of symmetry:

❖ Plane of Symmetry

❖ Centre of Symmetry

❖ Alternating axis of Symmetry

❖ Plane of Symmetry:

A molecule is said to possess as plane of symmetry when an imaginary plane passing through the center of molecule can divide it into two parts such that one is the exact mirror image of the other. If a molecule has plane of symmetry then the molecule and its mirror image are superimposable and hence molecule is optically inactive or achiral.

Few molecules with plane of symmetry are shown below:

```
          OH
           |
   H ——— C ——— COOH
- - - - - - | - - - - - - - - -  Plane of Symmetry
   H ——— C ——— COOH
           |
          OH
```

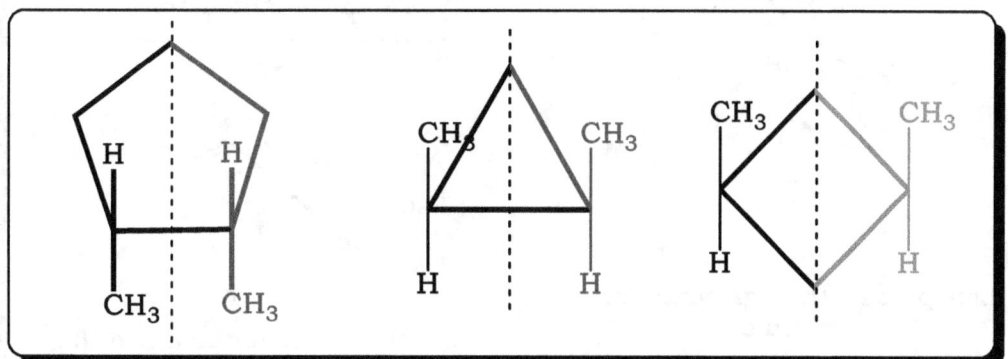

❖ **Centre of Symmetry:**

It is an imaginary point in the molecule from which the similar groups are at equidistant. Generally center of symmetry is observed in the even membered rings. The compounds having center of symmetry are optically inactive or achiral.

Following isomer of 1,3-dichloro-2,4-difluorocyclobutane has a center of Symmetry

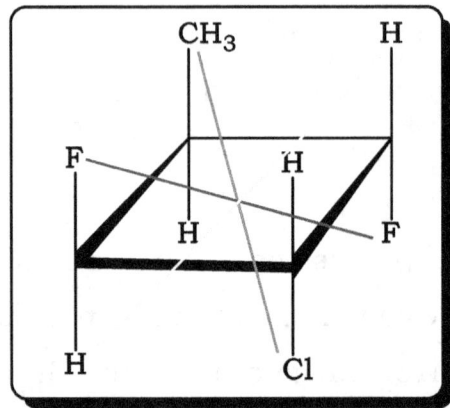

❖ **Alternating axis of Symmetry:**

It is a line about which the crystal may be rotated so that it represents the same appearance more than once during a complete revolution.

After rotating Through 180⁰

(a)
Line passing through centre of molecule

(b)

Plane of Symmetry

(c)
Mirror images of (b) (identical with a)

Many compound which posses alternating axis of symmetry are achiral

A molecule is said to have alternating axis symmetry, if an identical structure results when it is rotated around the axis by an angle of $2\Pi/n$ (n = number of fold of symmetry) and then reflected across the plane perpendicular to the axis. For example, 3,4-dibromo-3, 4-dimethyl hexane has alternating axis of symmetry.

Many compounds which possess alternating axis of symmetry are achiral. Thus a molecule that has a plane of symmetry, a center of symmetry and an alternating axis of symmetry is superimposable on its mirror image and is optically inactive and a molecule that has no element of symmetry is not superimposable with its mirror image and is optically active.

CONDITION OF OPTICAL ACTIVITY:

A chiral molecule has a center of chirality within it and it consist of a suitable atom substituted in a way so as to be non-superimposable on its mirror image. Thus chirality is the property of a molecule of being non superimposable on its mirror image. The most common feature which gives chirality to the molecule is a chiral center but not always. We can say that many but not all molecules that contain chiral center are chiral and many but not all chiral molecules contain chiral center.

Chirality is necessary and sufficient condition for exhibiting optical activity. The optical activity of sodium bromate, sodium iodate, quartz is lost when their lattice structure is destroyed by melting or dissolving in the water. This indicates that chirality is in the lattice structure.

Many organic compounds show optical activity even in solid, liquid, vapor or in solution form. This indicates that chirality is inherent in the molecule. Hence, they show optical activity even if their physical state is changed. That is they have molecular chirality.

ASYMMETRIC AND DISSYMMETRIC MOLECULES:

A disymmetric molecule lacks those elements of symmetry which preclude a mirror image relationship, whereas asymmetric molecule has no elements of symmetry at all.

Dissymmetric and asymmetric molecules are usually optically active. (A molecule having none of the elements of symmetry which preclude a mirror image relationship or having only an axis of symmetry is not superimposable with its mirror image and is called dissymmetric.)

DETERMINING WHETHER A MOLECULE IS CHIRAL OR NOT:

A foolproof method of determine whether a molecule is chiral or not, is to construct molecular models of the molecule and its mirror image relationship and look if these pass the test of superimposition. The chiral molecules are those in which object is non super imposable on its mirror image.

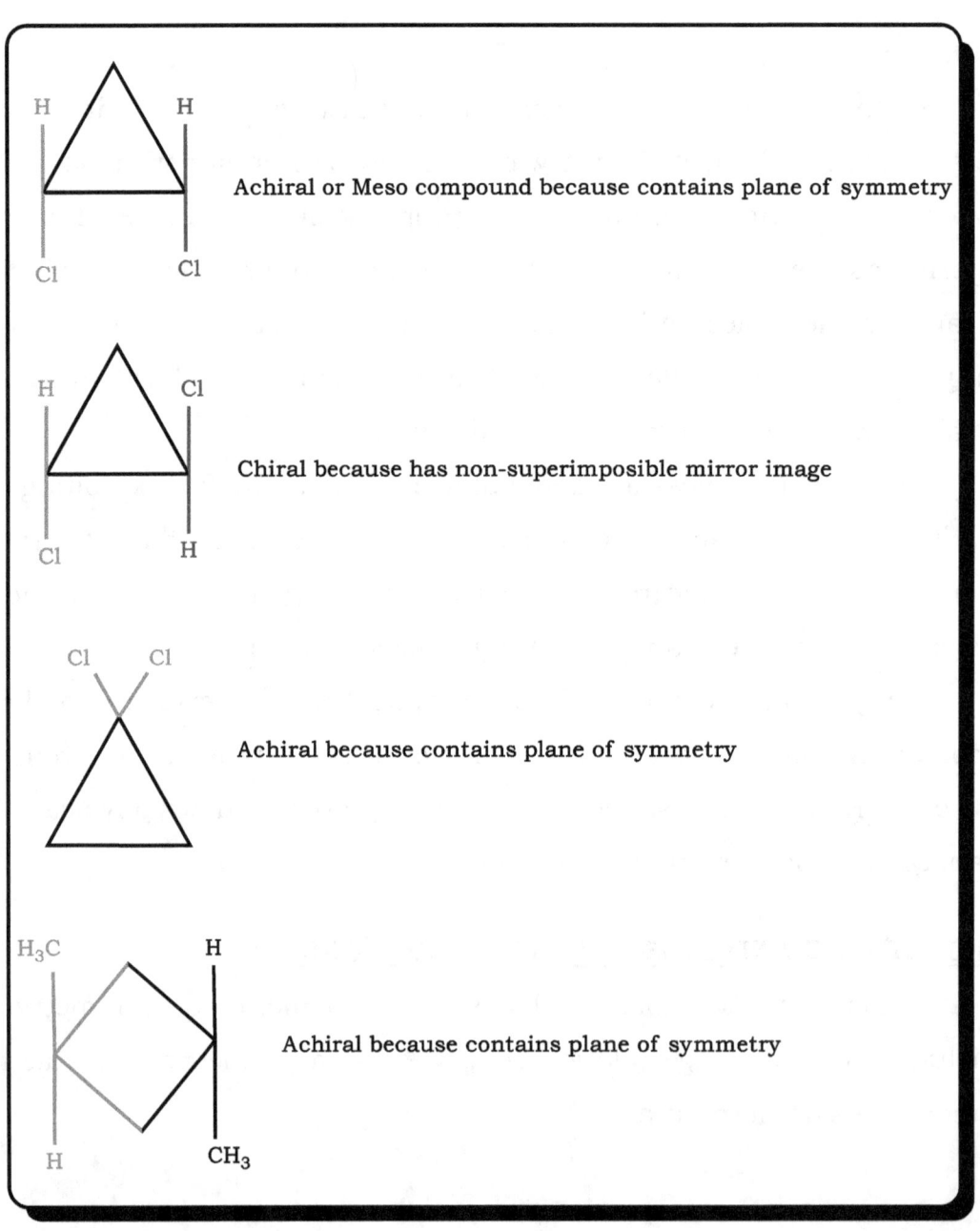

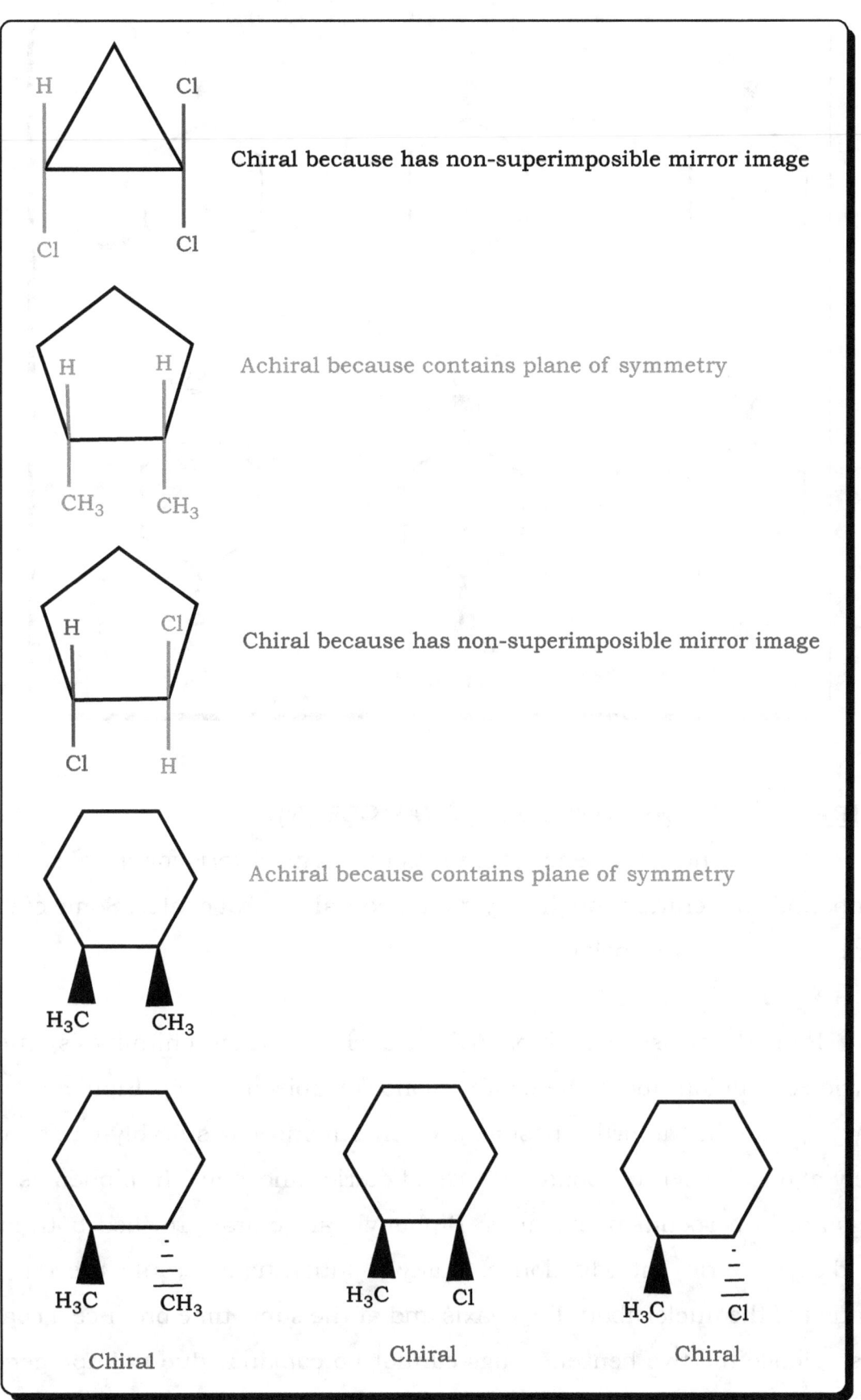

Chiral because has non-superimposible mirror image

Achiral because contains plane of symmetry

Chiral because has non-superimposible mirror image

Achiral because contains plane of symmetry

Chiral

Chiral

Chiral

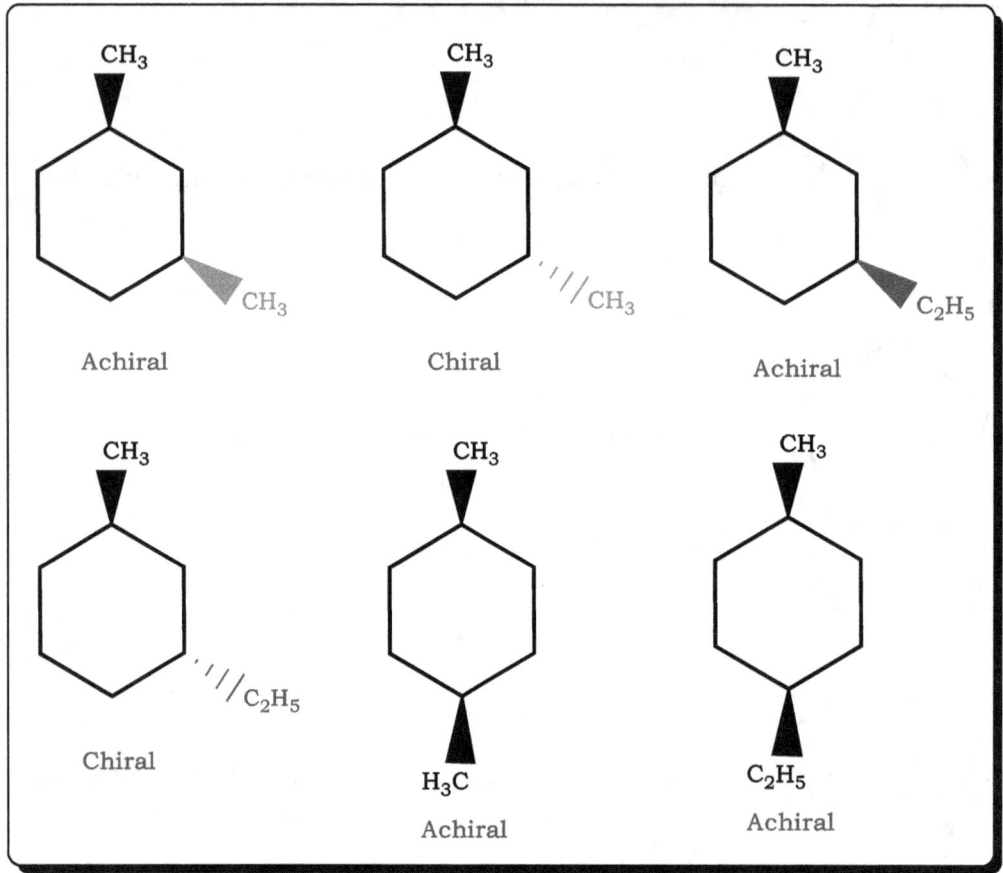

CHIRAL COMPOUND WITHOUT CHIRAL CENTRE:

The presence or absence of chiral center is no criteria for chirality. Many compounds are chiral though they do not contain chiral center. Some of such compounds are listed below:

❖ **BIPHENYLS:**

Properly substituted biphenyls are chiral. Their chirality is due to restricted rotation about the central bond (atropisomerism). Biphenyls with heavy groups in the ortho positions when substituted suitably can't rotate freely about the central bond because of steric hindrance. In biphenyls, two rings are in perpendicular planes. Biphenyls are chiral provided both sides are dissymmetric. Introduction of bulky o-substituents would prevent free rotation of the nuclei about the coaxis and at the same time produce a coaxial twist. Hence the two benzene rings cannot be coplanar due to impingement of the o-substituents and thus biphenyls become dissymmetric.

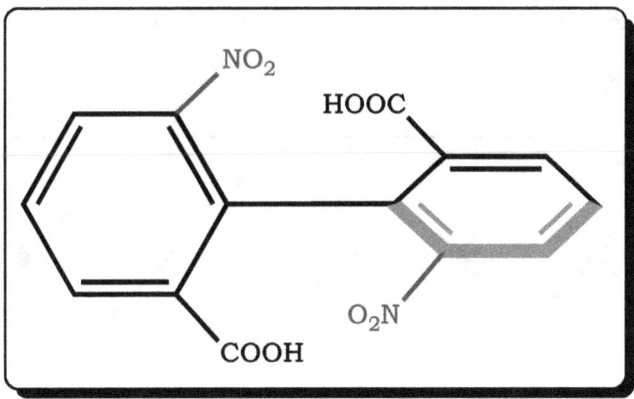

❖ ALLENES :

Suitably substituted allenes are chiral provided both the sides are dissymmetric. Following types of allenes will be dissymmetric provided that a ≠ b.

Enantiomeric allenes

Molecular dissymmetry is possible because the groups at one end of the allene molecule lie in a plane at right angles to those at the other end.

❖ SPIRANES:

Suitably substituted spiranes are chiral provided both sides are dissymmetric. In spiranes, the two are orthogonal as a consequence, groups attached to the ends of the system lie in planes which are mutually perpendicular.

Dissymmetric spiranes are obtained by attaching unequal substituents at each end of the system.

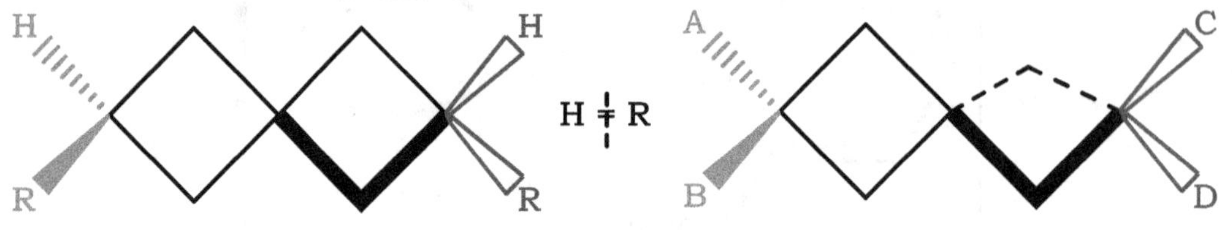

Dissymetric spiranes

❖ METHYLENE CYCLOALKANES:

Substituted methylene cycloalkanes are chiral provided both sides are dissymmetric. 4-Methylcyclohexyliden acetic acid was the first chiral compound of this type that was reported.

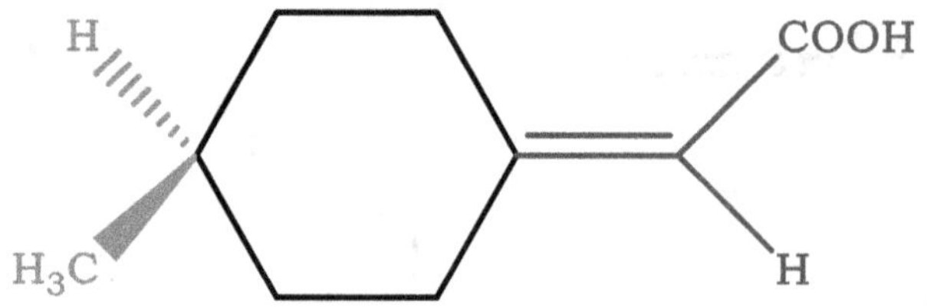

4-Methyl cyclohexyliden acetic acid

This substituted methylene cyclohexane is dissymmetric because the groups attached to the double bond lie in a plane at right angle to those attached to the 4-position of ring.

Another example of dissymmetric methylene cyclohexane is show below.

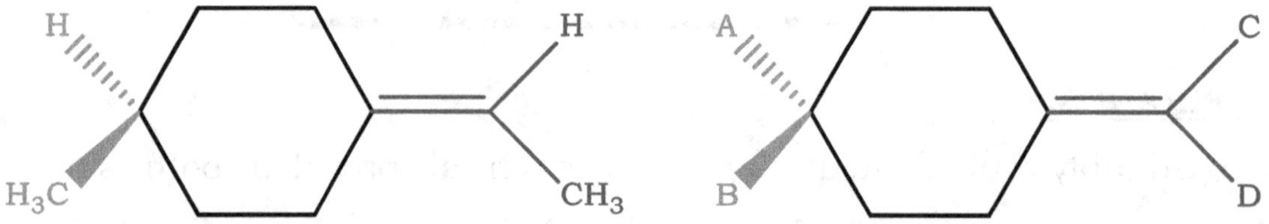

The stereoisomers which are otherwise same but differ in their action towards the plane polarized light are called optical isomers and this phenomenon is called **optical isomerism.** E.g. *d*-lactic acid and *l*-Lactic acid are optical isomers.

COOH   COOH

H   OH   HO   H

H_3C CH_3

d - Lactic acid   *l* - Lactic acid

$[\alpha] = +2.24^0$ $[\alpha] = -2.24^0$

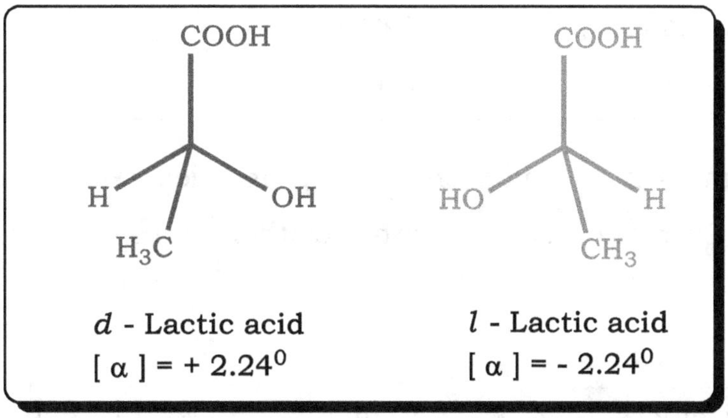

Similarly *d*-tartaric acid and *l*-tartaric acid are optical isomers.

ENANTIOMERS:

Stereoisomers that are mirror images of each other are called as enantiomers. For example, E.g. *d*-lactic acid and *l*-Lactic acid are enantiomers.

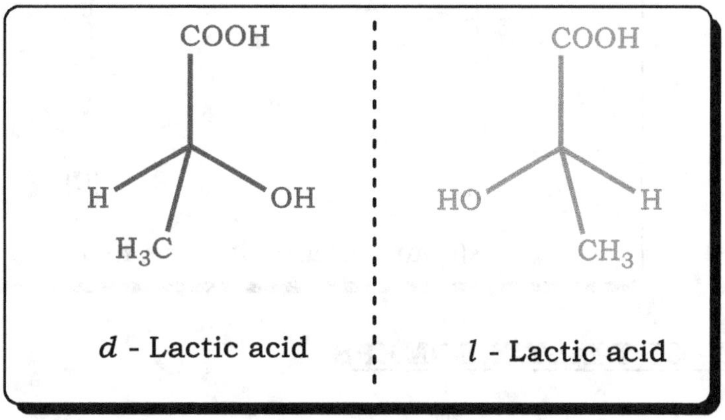

COOH   COOH

H   OH   HO   H

H_3C CH_3

d - Lactic acid   *l* - Lactic acid

PROPERTIES OF ENANTIOMERS:

❖ Enantiomers have identical physical properties such as boiling point, refractive index, relative density etc. but differ each other in their action on plane polarized light. If one of the enantiomer rotate the plane of plane polarized light to the right, the other will rotate to the left. However the extent of rotation is same.

The properties of two 2-methyl-1-butanols are shown below:

Property	(+)-2-Methyl-1-butanol	(-)-2-Methyl-1-butanol
Specific rotation	+5.90°	- 5.90°
Boiling point	128.9°C	128.9°C
Relative density	0.8193	0.8193
Refractive index	1.4107	1.4107

❖ Enantiomers have same chemical properties except the fact that they differ in the rate of reaction with the other optically active reagents (chiral probes).

DIASTEREOISOMERS:

Stereoisomers that are not mirror images of each other are called diastereoisomers. e.g.

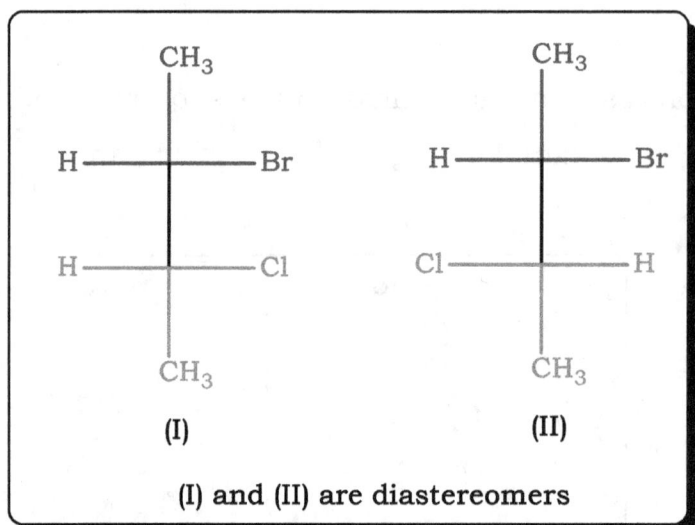

(I) and (II) are diastereomers

PROPERTIES OF DIASTEREOMERS:

❖ Diastereoisomers have different physical properties like melting point, boiling point, solubility in a given solvent, density, refractive indexes etc. Diastereoisomers differ in specific rotation; they may have the same or opposite signs of rotation, or some may be inactive.

❖ Diastereoisomers have different chemical properties.

EPIMERS:

Several sugars are closely related to each other and differ only by the stereochemistry at a single carbon atom. *Sugars which differ only by the stereochemistry at a single carbon atom are called epimers*. The carbon atom where the two sugars differ is generally stated and when it is not stated it is assumed to be C-2. For example, D(+)- glucose and D(+) – mannose are C-2 epimers. Similarly D-glucose and C-Galactose are C-4 epimers.

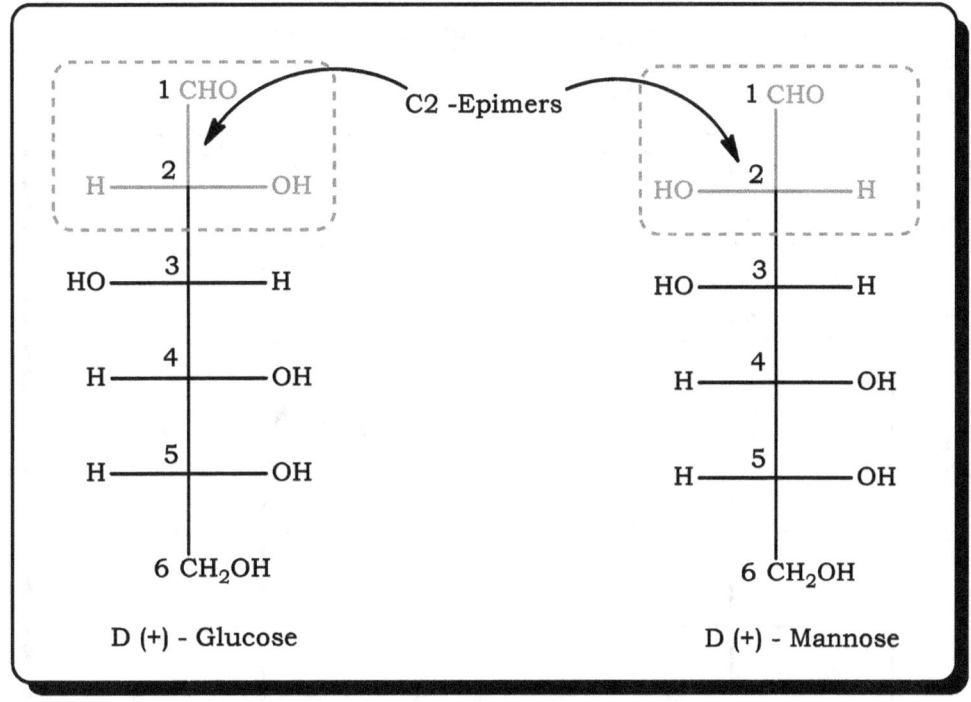

D (+) - Glucose   D (+) - Mannose

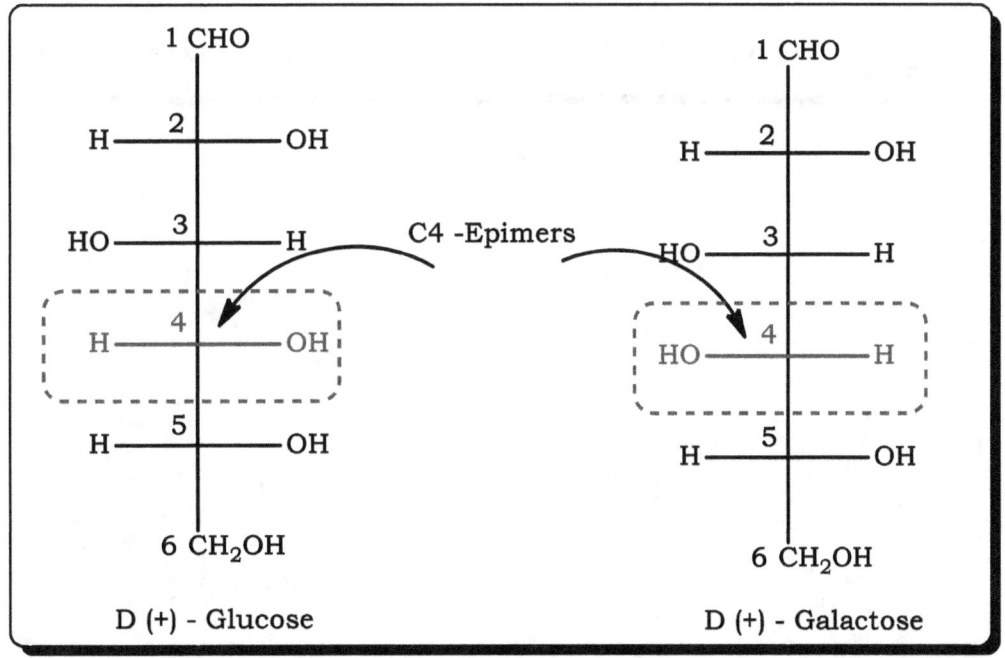

D (+) - Glucose   D (+) - Galactose

ANOMERS:

Glucose and other hexoses exist as an equilibrium mixture with their cyclic hemiacetal isomers in which the latter strongly predominates. The carbonyl carbon turns into a new stereo center on cyclization. Thus glucose has two cyclic forms which differ only in the stereochemistry at C-1 and the hemiacetal carbon is called as the *anomeric carbon*. Such isomers are called as anomers. The two anomers are commonly differentiated by the Greek letters α and β and thus in the case of glucose these diastereomers are termed α-D-glucose and β-D-glucose.

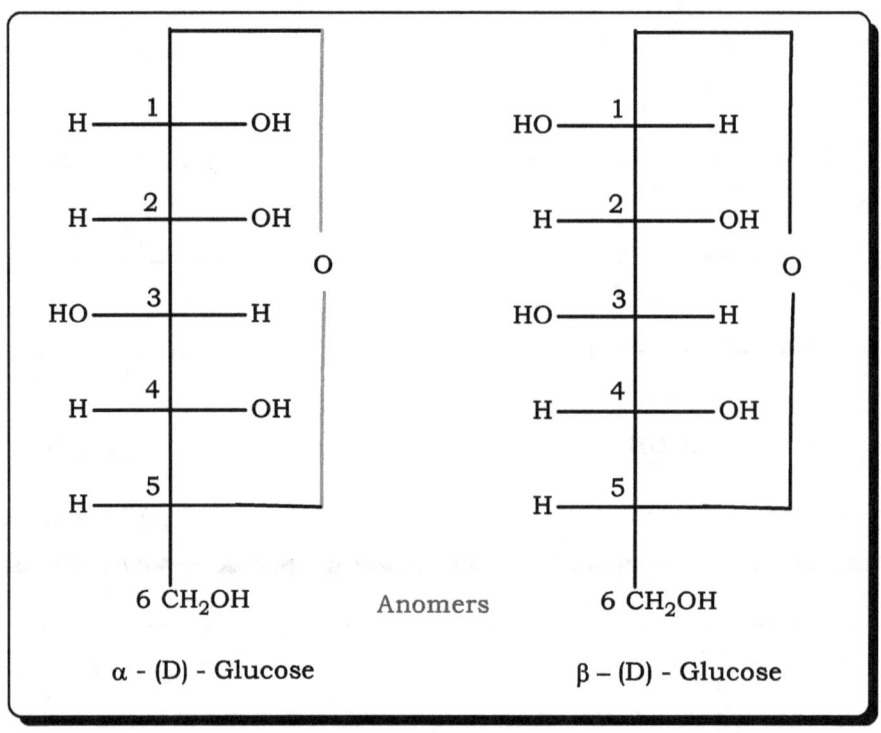

AN APPROACH TO THE CLASSIFICATION OF ISOMERS:

We can classify isomers by asking and answering a series of a questions:

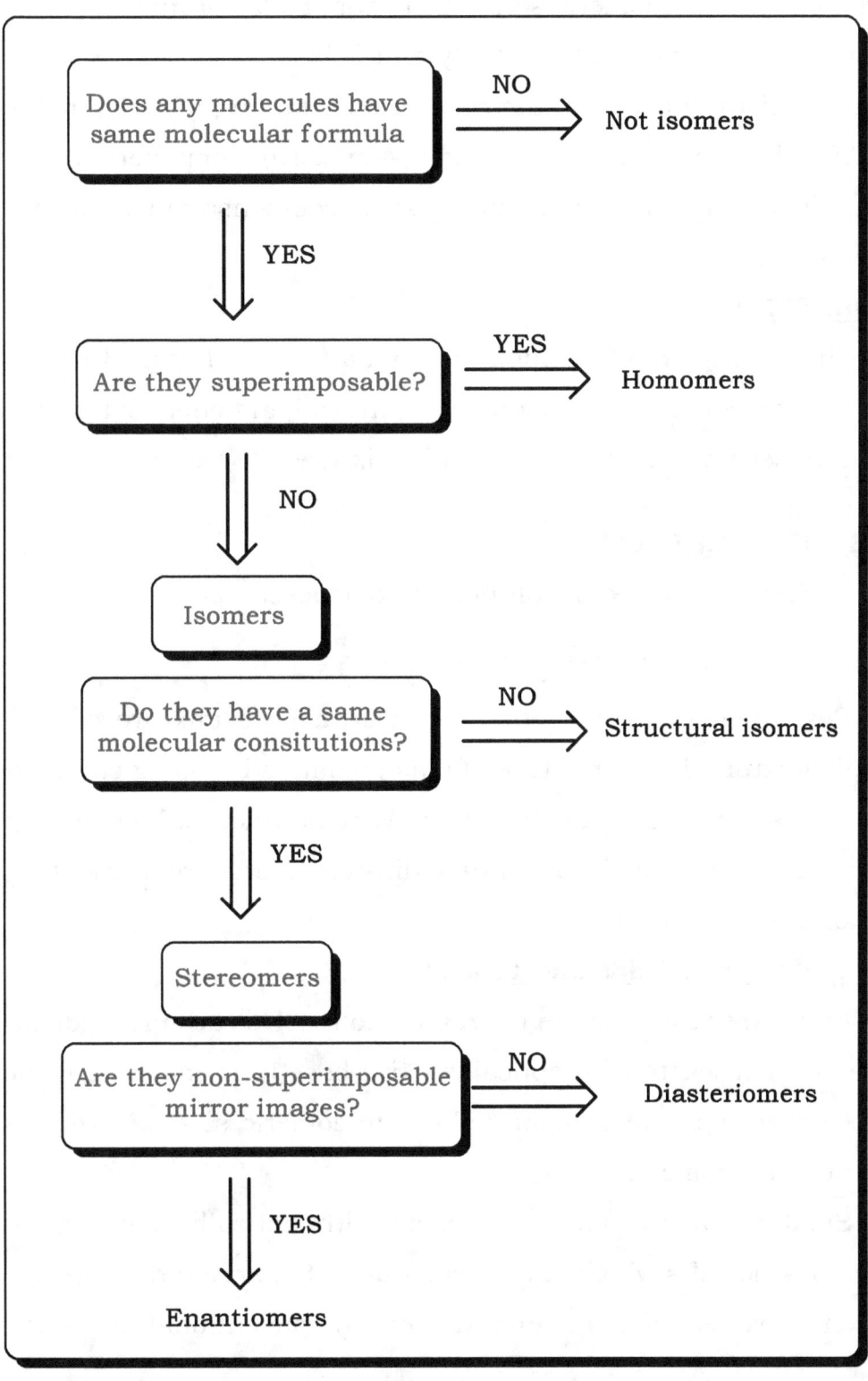

RACEMIC MIXTURE

Racemic modification or mixture is defined as mixture containing equimolar quantities of a pair enantiomers. It is represented as, (±) or *dl*. When a pair of enantiomers is mixed in equal molar proportion, the resulting mixture do not show optical activity and it is called *racemic mixture.* The optical inactivity is because, the right hand rotation (+) by dextro component is nullified by equal left hand rotation (-) by laevo component, since they are present in equal proportion. This type of compensation is called external compensation.

RESOLUTION:

The separation of enantiomers from a racemic is called *as resolution.* Since enantiomers have almost similar physical and chemical properties, it is difficult to separate them from racemic mixture.

METHODS OF RESOLUTION:

Various methods of resolution are discussed below.

1) USING CHIRAL PROBE:

Resolution is carried out by converting the mixture of enantiomers (racemic mixture) into a mixture of diastereomers by using chiral compound called as resolving agent. Since the resulting products will be diastereomeric, these can be separated. Separated diastereomers are converted back to enantiomers.

❖ **Resolution of Acids and Bases :**

Acid-base reactions are often used to resolve racemic acids and bases. A racemic acid reacts with optically active base to form diastereomeric salts which can be separated. Separated diastereomeric salts are then converted back into enantiomers.

Similarly, a racemic base reacts with optically active acid to form diastereomeric salts which can be separated. Separated diastereomeric salts are then converted back into enantiomers by conventional methods.

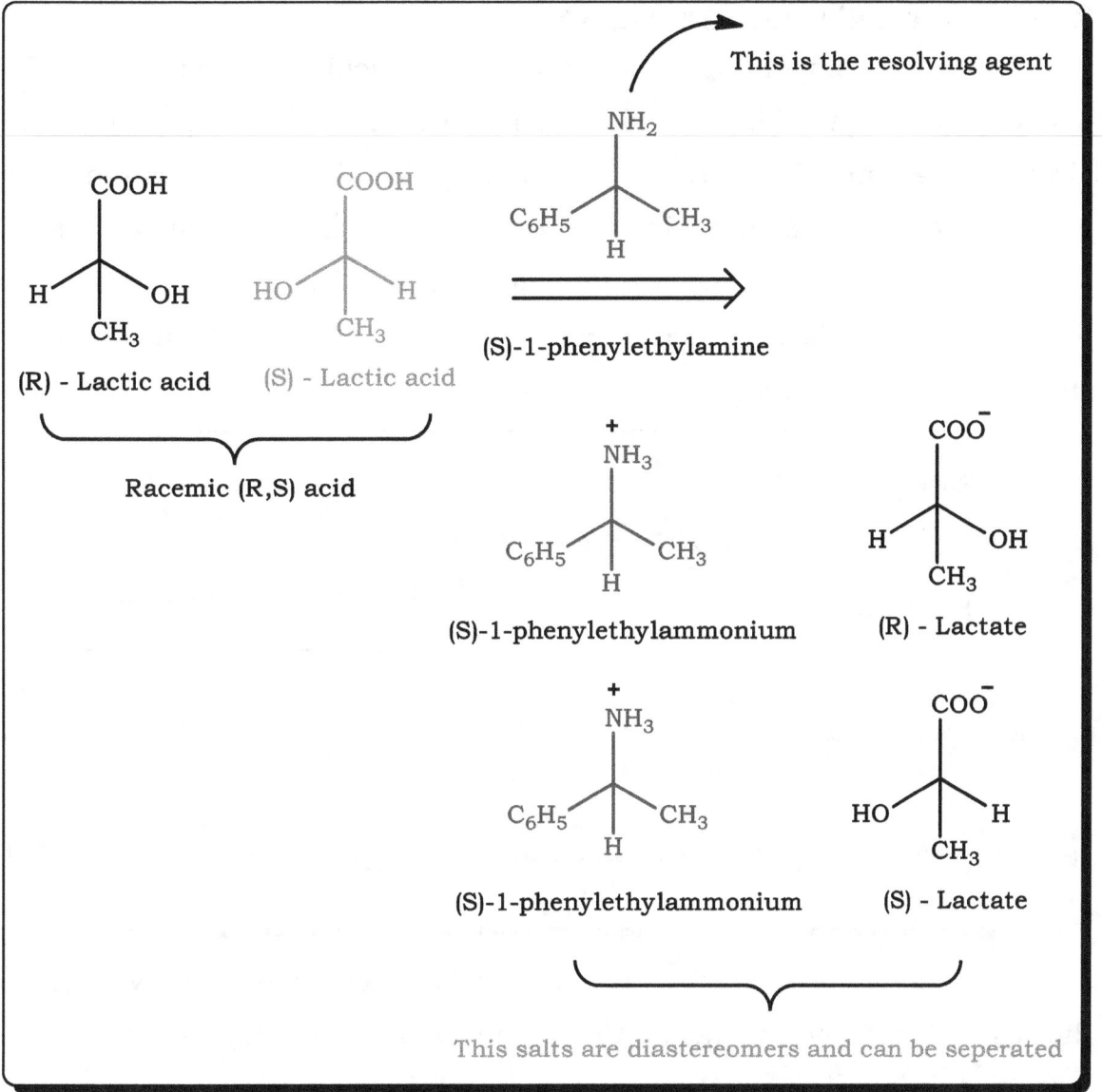

This is the resolving agent

(R) - Lactic acid   (S) - Lactic acid

Racemic (R,S) acid

(S)-1-phenylethylamine

(S)-1-phenylethylammonium   (R) - Lactate

(S)-1-phenylethylammonium   (S) - Lactate

This salts are diastereomers and can be seperated

(R) - acid   (S)-Amine   (S, S)-Salt   Seperate   (S, S)-Salt   +   (S, R)-Salt

    +   +

(S) - acid   (S, R)-Salt

                                                          ↓ HCl   ↓ HCl

                                                        (S)-acid   (R)-acid

                                                            +   +

                                                   (S)-Amm. chloride   (S)-Amm. chloride

The reslution of the racemic form of an organic acid

❖ Resolution of other compounds:

Racemic compounds which are neither acids nor bases are often resolved by first attaching an acidic handle. Consider an example of racemic alcohol. A racemic alcohol reacts with cyclic anhydride to form both an ester and an acid. The racemic acid thus produced is separated via diastereomeric salts and converted back to the enantiomeric acids. The acidic handle is then removed by hydrolysis of the ester group and the separated alcohol enantiomers are obtained.

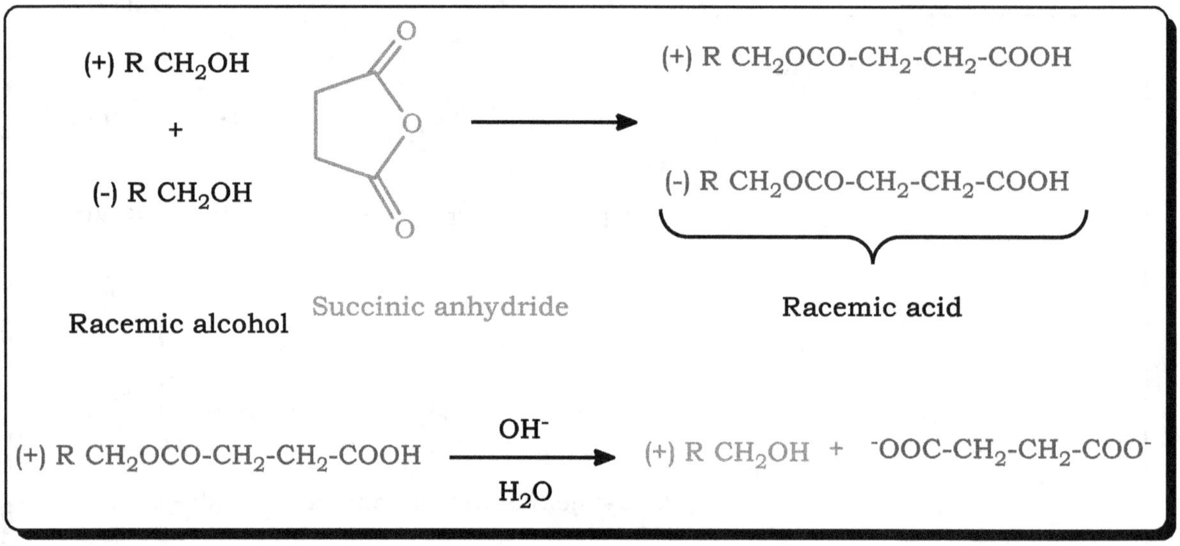

Various naturally occurring alkaloids (optically active) are available for the resolution of racemic acids. Common examples are strychnine, brucine, morphine, ephedrine, cinchonine, quinine and quinidine. One of the most important synthetic basic resolving agent is α-phenyl ethyl amine. Among the acids which are used in the resolution of racemic bases are camphoric acid, camphor-10-sulphonic acid, methyoxy acetic and pyrrolidine-5-carboxylic acid.

2) CHROMATOGRAPHIC TECHNIQUES:

The resolution by using chromatography depends on the difference in the rates of reaction of the two enantiomers with the chiral probe material on the column packing.

The method involves preparation of column of solid optically active compounds like tartaric acid, sucrose etc. The enantiomeric components of the racemic mixture form diastereomeric complexes with the chiral material used on column. (One enantiomer passes through the chromatographic column faster than the other due to varied rate with the chiral material).

The racemic mandelic acid has been resolved by column chromatography on starch. Racemic p-phenylenebisiminocamphor has been resolved on lactose column. Racemic Troger's base has been resolved on lactose column.

3) ENZYMATIC RESOLUTION:

Enzymes are complex optically active protenoid catalysts which are produced by living organism. A fundamental property of enzymatic reactions is high degree of stereo selectivity due to the asymmetric nature of enzymes. Enzymes show different rates or reaction with the two enantiomers. For example, a certain bacterium digest only one enantiomer and not the other. This method has limiting value since one of the enantiomer is destroyed during resolution.

For example, Penicillium glaucum destroys ammonium (+) tartrate more rapidly than (-) tartrate in a dilute solution of ammonium (±) tartrates.

Racemic α-amino acids are resolved by using swine kidney acylase.

4) MECHANICAL SEPARATION:

Enantiomers of very few substances can be crystallized into asymmetric crystals. Since the appearance of these crystals are different, a trained crystallographer can separate them with tweezers.

The method was first time used by Pasteur. The mechanical separation is an attractive method for effective resolution. However, difficulties arise due to the fact that very few substances can be crystallized into asymmetric crystals. Sodium ammonium tartrate is one of this rare group of compounds and even in this case the crystallization must be carried out below 27°C.

5) DIFFERENTIAL REACTIVITY:

Since enantiomers react with chiral compounds at different rates, it may be possible to effect a partial separation by stopping the reaction before it goes to completion.

The R, S convention is governed by three rules which are quite logical and easy to remember; these rules set a priority sequence for the four substituents, A, B, D, E, about an asymmetric carbon.

Rule-1: The priorities assigned to the four substituents on an asymmetric carbon depend upon the atomic number of the attached atom. Greater the atomic number of atom, greater is the priority. *[For example, the halogens have the priority I > Br > C1 > F; other obvious priorities would be C1 > O > N > C.]*

Let us now consider the configurational assignment for bromochlorofluromethane, which has the atomic priorities Br > C1 > F > H. Visualize the molecule in such a way that the atom with lowest priority projects behind the paper; the remaining three substituents are arranged in order of priority either clockwise or counterclockwise as follows

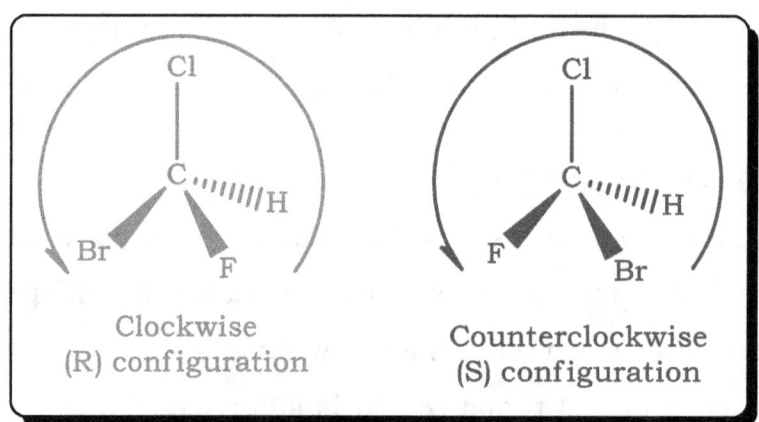

Clockwise
(R) configuration

Counterclockwise
(S) configuration

The clockwise direction is assigned the R configuration (Latin: rectus, right) and the counterclockwise direction is assigned the S configuration (Latin: sinister, left).

Rule-2: When two or more substituents on the asymmetric carbon have atoms with the same atomic number, then Rule 2 applies; precedence is given to the substituent with the highest atomic number in the second atom from the asymmetric center.

For example, with sec-butyl alcohol, two of the adjacent atoms are C. However, the C of –CH$_2$-CH$_3$ is bonded to C, whereas the C of -CH$_3$ is bonded only to H; therefore, -CH$_2$-CH$_3$ takes priority order for sec-butyl alcohol is OH > C$_2$H$_5$ > CH$_3$ > H.

Sec-Butyl alcohol

Rule-3: If the second atom from the asymmetric center has an identical atomic number in two groups, then the total of the atomic numbers attached to the first carbon are considered. With this rule a double bond, -C = A, is treated as –CA$_2$. For example, -C = O would be treated as -CO$_2$. Similarly a triple bond, CA, would be treated as –CA$_3$.

With this rule, the carboxyl (-COOH) with combined atomic numbers of 48 would take precedence over an aldehyde (-CH=O) with atomic numbers of 33; similarly –CH=O would take precedence over –CH$_2$OH (combined atomic numbers = 18). Thus in the case of glyceraldehyde, the priorities are

- OH > -CH=O > CH$_2$OH > H.

The phenyl substituent is treated as –C-C$_3$ with combined atomic numbers of 36; thus phenyl would take priority over isopropyl, -CH(CH$_3$)$_2$ with combined atomic numbers of 25.

By applying these rules to some common substituents, one obtains the following sequence (group of highest priority first): I, Br, CI, SO$_2$R, SOR, SR, SH, F, OCOR, OR, OH, NO$_2$, NHCOR, NR$_2$, NHR, NH$_2$, CCI$_3$, COCI, CO$_2$R, CONH$_2$, COR, CHO, CR$_2$OH, CHOHR, CH$_2$OH, CR$_3$,C$_6$H$_5$, CHR$_2$, CH$_2$R, CH$_3$, D, H.

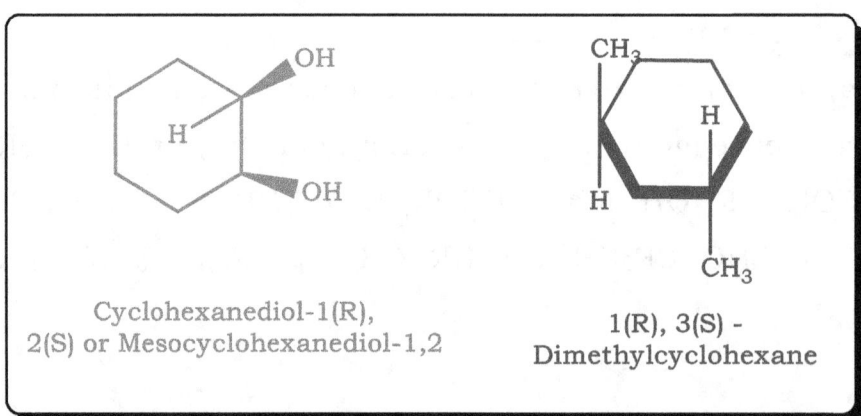

R-Glyceraldehyde

CHO

C

HO ----- H

CH_2OH

S-Alanine

$COOH$

C

H_3C ----- H

NH_2

If more than one asymmetric carbon is present, then each configuration is specified, along with the position number of the asymmetric carbon.

HO H $COOH$

C

C

$HOOC$ OH

H

2 (R),(R)-Tartaric acid

$COOH$

H — C — OH

HO — C — H

$COOH$

2 (R), 3(R)-Dihydroxy succinic acid

CHO

HO — C — H

H — C — OH

HO — C — H

CH_2OH

2 (R), 3(R), 4(R), 5-Tetrahydroxy-Pentanal

OH

H

OH

Cyclohexanediol-1(R), 2(S) or Mesocyclohexanediol-1,2

CH_3

H

H

CH_3

1(R), 3(S) - Dimethylcyclohexane

Q-1: Find out the enantiomer pair amongst following compounds.

	COOH			COOH			COOH	
H	—	OH	H	—	OH	HO	—	H
H	—	OH	HO	—	H	H	—	OH
	COOH			COOH			COOH	
	(A)			(B)			(C)	

Solution: B and C are enantiomeric pair.

Q-2: Find out relationship between (i) and (ii).

	CH₃			CH₃	
H	—	OH	H	—	OH
H	—	OH	HO	—	H
H	—	OH	H	—	OH
	CH₃			CH₃	
	(I)			(II)	

Solution: i and ii are diastereomers.

Q-3: Choose the correct relation of (ii) and (iii) with (i).

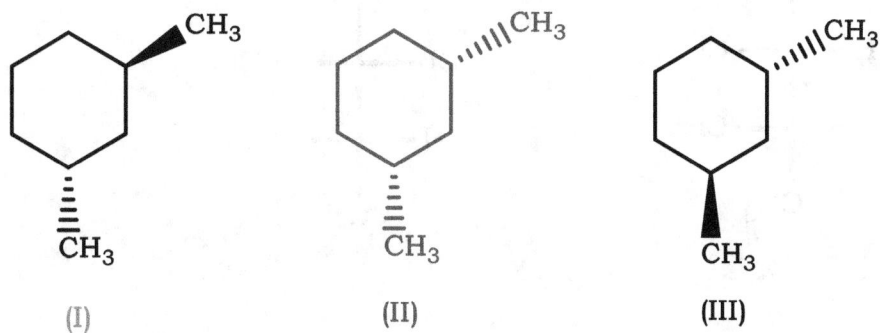

(I) (II) (III)

Solution: Compound (ii) is a diastereomer of (i) whereas (iii) is enantiomer of (i).

Q-4: Identify the relationship between following pairs of structures.

(A) 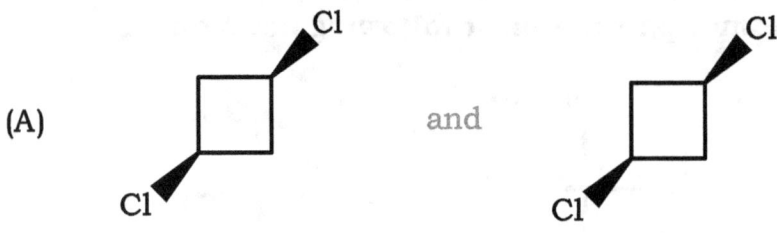 and

(B)

$$CH_3$$
H —|— Cl
H —|— Br
$$CH_3$$

and

$$CH_3$$
Cl —|— H
Br —|— H
$$CH_3$$

(C)

Cl ⟍ ⟋ Cl
Cl ⟍ ⟋ Cl

and

Cl ⟍ ⟋ Cl
Cl ⟍ ⟋ Cl

(D)

$$CH_3$$ (on cyclobutane)

and

(cyclopentane)

(D)

$$CH_3$$
H —|— Cl
H —|— Cl
$$CH_3$$

and

$$CH_3$$
Cl —|— H
H —|— Cl
$$CH_3$$

Solution:

a) Diastereomers   b) Enantiomers   c) Same   d) Constitutional isomers

e) Diastereomers

Q-5: How many stereoisomers are possible for cholesterol?

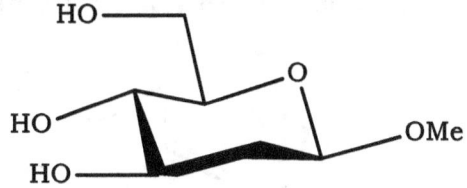

Solution: Number of stereoisomers = 2^8 =256.

Q-6: Menthol has three asymmetric centers. How many epimers and diastereoisomers are possible for it?

Solution: 3 Epimers and 6 diastereoisomers.

Q-7: Methyl-D-glusoside has five asymmetric centers how many epimers (E) and diastereoisomer (D) are possible for this molecule?

Solution: 5 Epimers and 30 Diastereoisomers.

Q-8: Natural cholesterol on hydrogenation affords cholesterol as one of the products as this compound has nine asymmetric centers, it can exists as 2^9 i.e. 512 possible isomers. The number of diastereomers of cholesterol are

Solution: 510.

STEREOCHEMISTRY

QUESTIONS ASKED IN SET EAMINATIONS

1) Natural cholesterol has 8 chiral centers. It can therefore exist as 2^8 i.e. 256 possible isomer. The number of diastereomers of natural cholesterol are

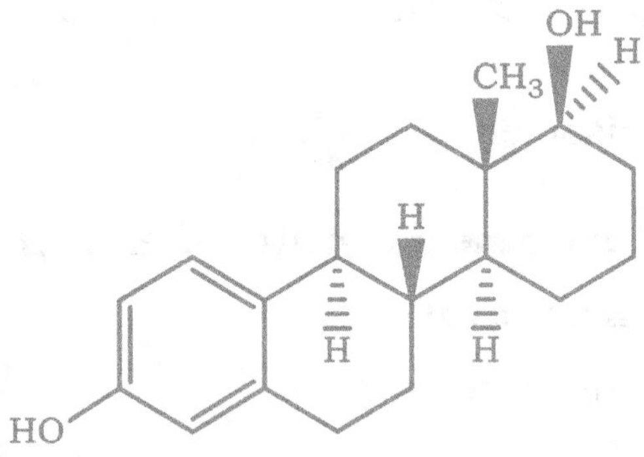

Cholesterol

A) 64   B) 128   C) 254   D) 255

2) **The configuration of the compound shown below is**

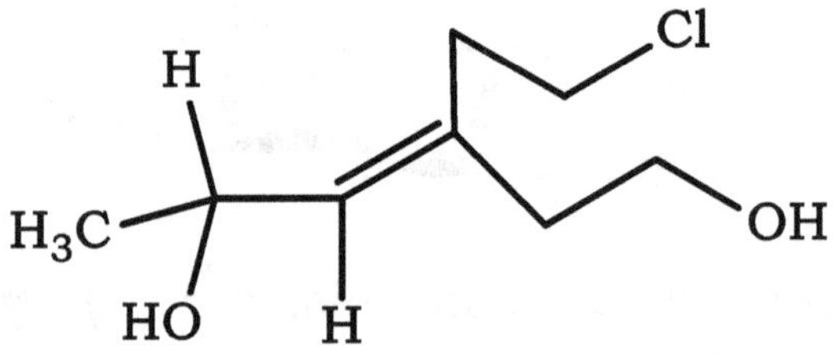

A) 2 R, 3 E   B) 2 S, 3 E   C) 2 R, 3Z   D) 2 S, 3 Z

3) For the two pairs of compounds given below, choose the correct statement

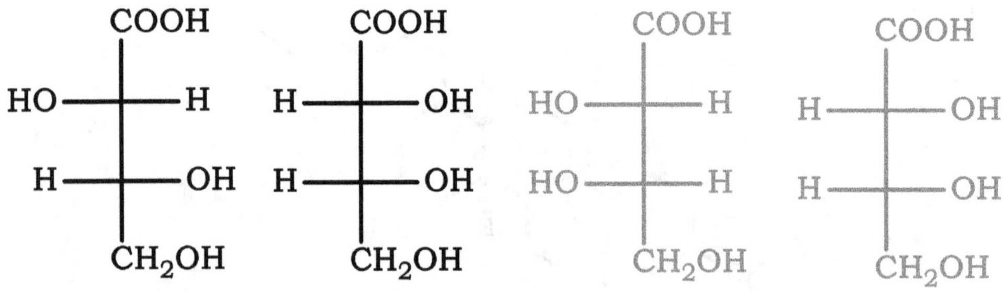

(I)   (II)

A) Pair (i) is enantiomeric while pair (ii) is diastereomeric

B) Pair (i) is diastereomeric while pair (ii) is enantiomeric

C) Both pairs are diastereomeric

D) Both pairs are enantiomeric.

4) **In the resolution of 1-phenylethylamine using [S] – (-) maleic acid, the compound obtained by recrystallization of the mixture of the diastereomeric salt is [R] - 1 – phenyl ethyl ammonium – [S] – maleate. The other component of the mixture is more soluble & remains in solution. The configuration of the more soluble salt is-**
A) [S] – 1 – phenylethylammonium – [S] - maleate
B) [S] – 1 – phenylethylammonium – [R] - maleate
C) [R] – 1 – phenylethylammonium – [R] - maleate
D) [R] – 1 – phenylethylammonium – [S] – maleate

5) The chiral molecule among the following are,

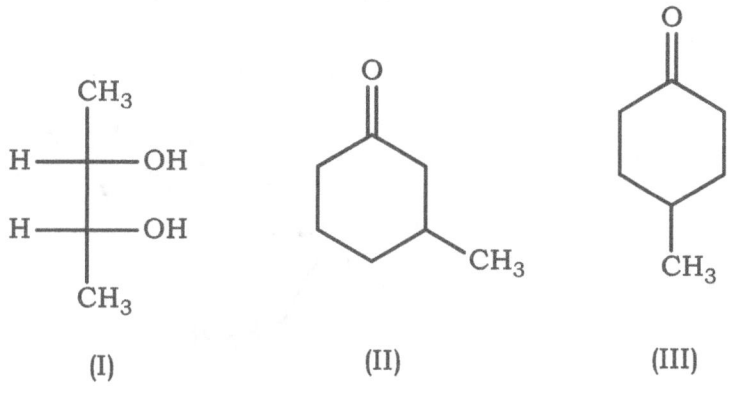

(I) (II) (III)

A) All three   B) (ii) & (iii)   C) (i) & (ii)   D) Only (ii)

6) **The configuration of the compound shown below is:**

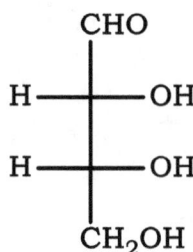

A) 2S, 3S   B) 2S, 3R   C) 2R, 3S   D) 2R, 3R

7) Choose the correct statement regarding the compounds shown below:

CH_3 structure (I):

$$Br \longrightarrow OH$$
with CH_3 top, Br bottom, H and CH_3 middle

Structure (II): alkene with H, CH_3 on top carbon and HOOC, H on bottom carbon

(I)   (II)

A) Both can be resolved   B) Both cannot be resolved

C) Compound (i) can be resolved but compound (II) cannot be resolved as it has a plane of symmetry passing through the olefin carbons & also through all three substituents.

D) Compound (II) can be resolved but compound (I) cannot be resolved as it has plane of symmetry after rotation of lower carbon by 120°.

8) The absolute configuration of the comp given is:

Structure: alkene with HOH_2C and C_2H_5 on top carbon; H_3C and C (bearing OH, H, CH_2OH) on bottom carbon

A) 2S, 3E   B) 2S, 3Z   C) 2R, 3E   D) 2R, 3Z

9) Choose the correct relation of (ii) & (iii) with (i)

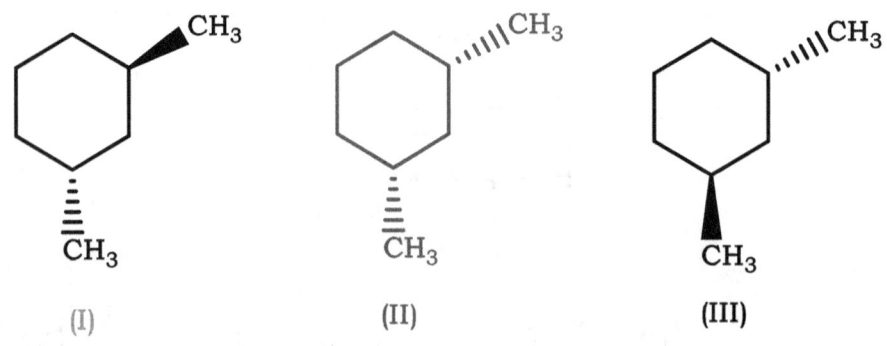

(I)   (II)   (III)

A) Both (ii) and (iii) are enantiomers of (i)

B) Both (ii) and (iii) are diastereomers of (i)

C) Compound (ii) is a diastereomer of (i) while (iii) is an enantiomer of (i).

D) Compound (ii) is an enantiomer of (i) while (iii) is a diastereomer of (i).

10) **The geometry of the two double bonds in the given molecule is**

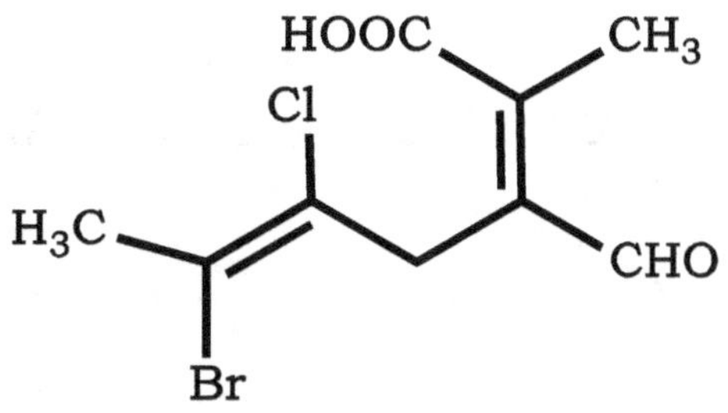

A) 2Z, 5Z   B) 2Z, 5E   C) 2E, 5E   D) 2E, 5Z

11) Chair form of cyclohexane has got these many gauche interactions:

A) 6   B) 2

C) 12   D) 0

12) Menthol has three asymmetric centers. How many epimers (E) and diastereomers (D) are possible for it?

A) E3; D8   B) E3 ; D6

C) E1 ; D8   D) E2 ; D6

13) **In the compound shown below the geometry of the two double bond is:**

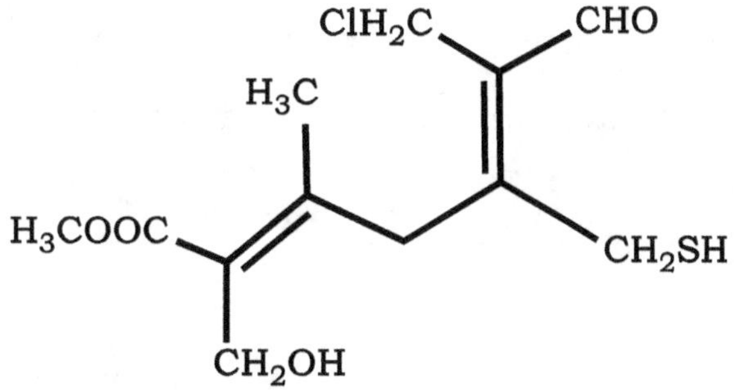

A) 2Z, 5Z   B) 2Z, 5E   C) 2E, 5E   D) 2E, 5Z

14) β-methyl-D-glucoside has five asymmetric centers. How many epimers (E) and diastereoisomers (D) are possible for this molecule?

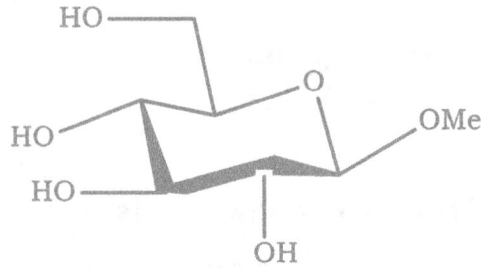

A) E-5 ; D-30   B) E-5 ; D-31   C) E-4 ; D-32   D) E-4 ; D-30

15) The chair form of cyclohexane has lower enthalpy than boat form because the boat has

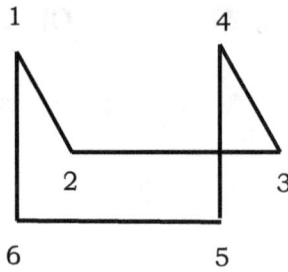

A)   The bowsprit flag pole interaction.

B)   Eclipsing interaction at C_1/C_2 and C_4/C_5

C)   Both bowsprit flagpole interaction & eclipsing interactions C_1/C_2 and C_4/C_5

D)   Both bowsprit flagpole interaction & eclipsing interactions at C_2/C_3 and C_5/C_6.

16) The specific rotation of a molecule in a 5 cm polarimeter tube is +2°. What will be its rotation in 10 cm. tube?

A) -4°   B) 20°

C) +4°   D) -2°

17) Select the appropriate stereochemical relationship among the following two compounds.

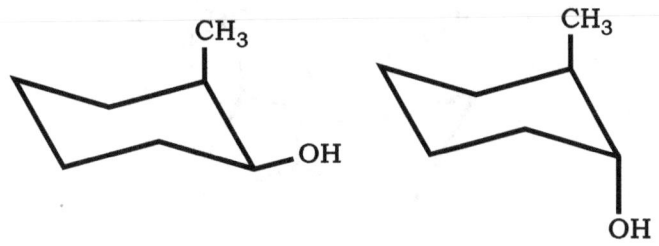

A) Structural isomers   B) Cis-trans isomers

C) Conformational isomers   D) R & S isomers

18) The compound shown below on treatment with base will

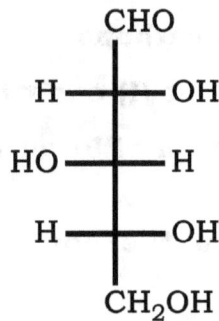

A) Undergo equilibrium   B) Be recovered unchanged

C) Undergo epimerization at C-5   D) Undergo epimerization at C-7.

19) Choose the correct configurational nomenclature for the following molecule:

$$
\begin{array}{c}
\text{CHO} \\
\text{H} \!-\!\!-\!\!-\! \text{OH} \\
\text{HO} \!-\!\!-\!\!-\! \text{H} \\
\text{H} \!-\!\!-\!\!-\! \text{OH} \\
\text{CH}_2\text{OH}
\end{array}
$$

A) 2R, 3S, 4S   B) 2S, 3R, 4S   C) 2R, 3S, 4R   D) 2R, 3R, 4S

20) Axial methyl groups is less stable 1.8 kcal/mol than the equatorial counterpart due to 1,3-diaxial interactions between CH_3 and H

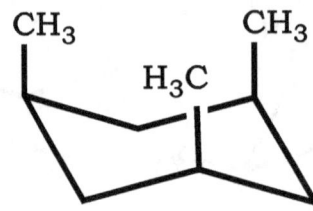

For the given molecule, the energy estimated will be,

A) 3.6 kcal/mol

B) More than 3.6 kcal/mol

C) 5.4 kcal/mol

D) More than 5.4 kcal/mol

21) The correct statement regarding the compounds shown is

COOH

H —— OH

HOOC —— H

OH

(I)

H_3C CH_2CH_3

H   COOH

(II)

A) Both can be resolved

B) Both cannot be resolved

C) Compound (I) can be but (II) can't be resolved as it has plane of symmetry passing through both olefinic carbons and all four substituent on these carbons.

D) Compound II can be but (I) can't be resolved as it has plane of symmetry after rotation of lower carbon by 120°

22) In the compound shown the configuration of the three double bonds is

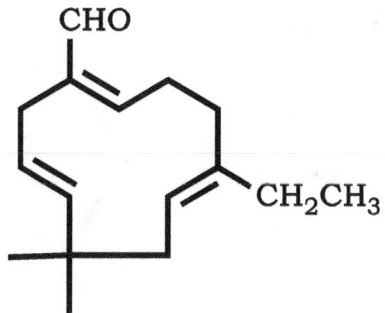

A) 2E, 6E, 9E   B) 2Z, 6Z, 9E

C) 2E, 3Z, 9E   D) 2Z, 6Z, 9Z

23) Predict the relation between following two pairs

CH₃		CH₃		CH₃		COOH	
HO ── H		H ── OH		HO ── H		HO ── H	
HO ── H		H ── OH		HO ── H		H ── OH	
Ph		Ph		Ph		Ph	

(I)   (II)

A) Both pairs are enantiomer

B) Both pairs are diastereomeric

C) Pair (i) is enantiomer while pair (ii) is diastereomeric

D) Pair (i) is diastereomeric while pair (ii) is enantiomer

24) The configuration of the compound shown below is

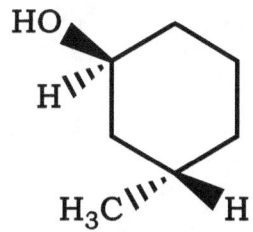

A) 1R, 3R   B) 1S, 3S   C) 1R, 3S   D) 1S, 3R

25) The Fischer projection of

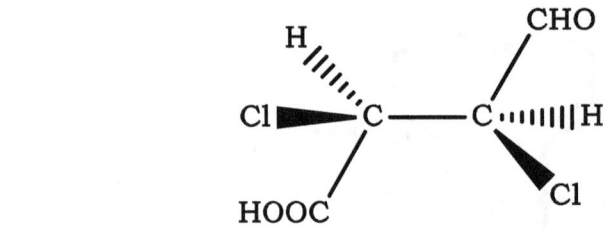

A)

$$\begin{array}{c} CHO \\ Cl \overline{} H \\ Cl \overline{} H \\ COOH \end{array}$$

B)

$$\begin{array}{c} CHO \\ Cl \overline{} H \\ H \overline{} Cl \\ COOH \end{array}$$

C)

$$\begin{array}{c} COOH \\ H \overline{} Cl \\ H \overline{} Cl \\ CHO \end{array}$$

D)

$$\begin{array}{c} CHO \\ H \overline{} Cl \\ H \overline{} Cl \\ COOH \end{array}$$

26) (S)-2chloro-2-ethyl pentanoic acid is represented as:

A)

$$\begin{array}{c} Pr \\ Cl \overline{} COOH \\ Et \end{array}$$

B)

$$\begin{array}{c} Pr \\ Et \overline{} Cl \\ COOH \end{array}$$

C)

$$\begin{array}{c} Et \\ Cl \overline{} Pr \\ COOH \end{array}$$

D)

$$\begin{array}{c} Et \\ Pr \overline{} COOH \\ Cl \end{array}$$

27) The correct configuration of the double bonds at C_2, C_4 & C_6 in the following compound is given by

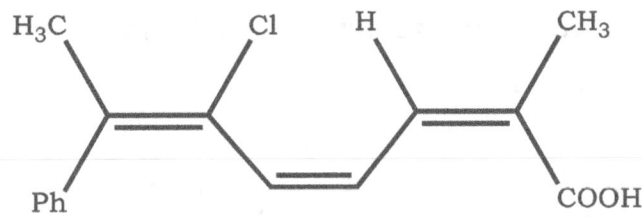

A) 2(Z), 4(Z), 6(E)   B) 2(Z), 4(E), 6(Z)

C) 2(Z), 4(E), 6(E)   D) 2(E), 4(Z), 6(E)

28)

$$
\begin{array}{c}
CH_3 \\
H {-}{\mid}{-} Cl \\
Ph {-}{\mid}{-} Cl \\
CH_3
\end{array}
$$

can be represented by:

A) 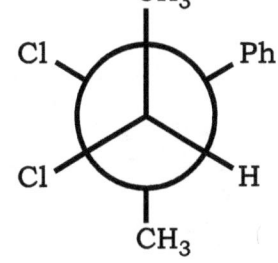   B)

C) 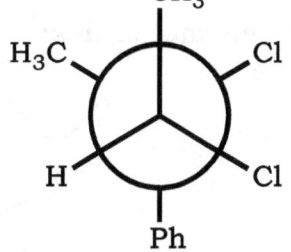   D)

29) Which one of the following is optically in active?

A)   B)   C)   D)

30) **How many stereoisomers are possible for estradiol having the following structure?**

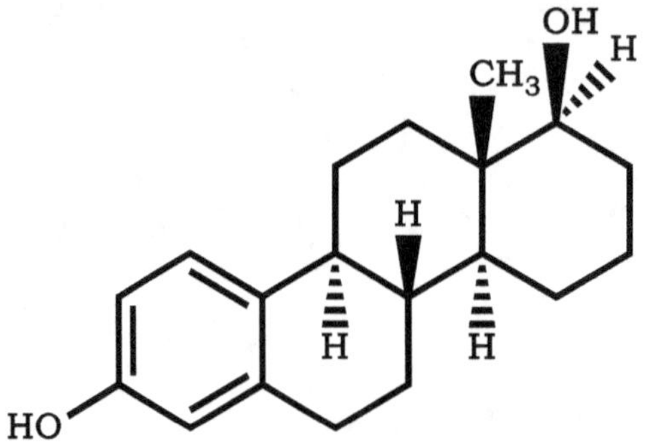

A) 5   B) 32   C) 16   D) 10

31) When a given optical active compound A is hydrolyzed with aqueous acid, it would be give:

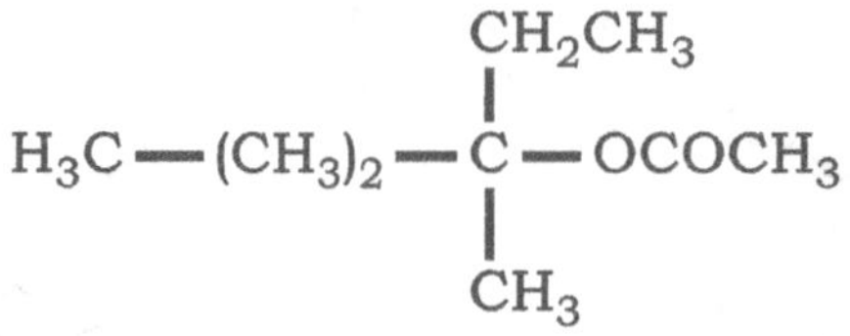

$$H_3C-(CH_3)_2-\underset{\underset{CH_3}{|}}{\overset{\overset{CH_2CH_3}{|}}{C}}-OCOCH_3$$

A) (-) Alcohol   B) (+) Alcohol   C) (+ & -) Alcohol   D) n-octanol

32) **Choose the correct geometrical relationship between various groups in the given molecules:**

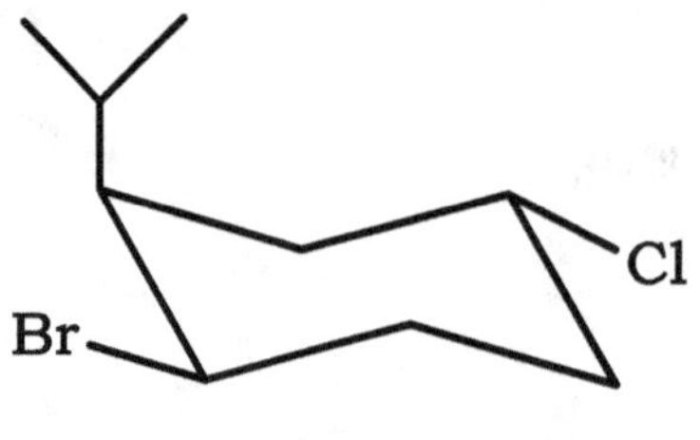

A) Cl and Br trans; Cl & i-Pr trans; Br & i-Pr trans

B) Br & i-Pr cis; Cl & i-Pr- trans; Br & Cl trans

C) Cl & Br – cis; Cl & i-Pr (trans); Br & i-Pr cis

D) Cl & Br – trans; Cl & i-Pr cis; Br & i-Pr cis

33) Select the correct geometry of the two double bonds in the following molecule:

A) 2Z, 5Z   B) 2E, 5E   C) 2E, 5Z   D) 2Z, 5E

34) The specific rotation of a compound measured in a 5 cm polarimeter is +2° at the given concentration, temperature & wavelength. Its specific rotation measured in 10 cm tube under similar condition will be

A) +4°   B) -4°   C) -2°   D) +2°